Basic Relationships of Gas Chromatography

Leslie S. Ettre

Department of Chemical Engineering
Yale University
New Haven, Connecticut

John V. Hinshaw

The Perkin-Elmer Corporation
Norwalk, Connecticut

1993 Edition

ADVANSTAR
DATA

Cleveland, Ohio

Printed in the United States of America

10 9 8 7 6 5 4 3 2 1

ISBN 0-929870-18-2 (Softcover)

Library of Congress Catalog Card Number: 93-71589

Published by ADVANSTAR Communications

ADVANSTAR Communications is a U.S. business information company that publishes magazines and journals, produces expositions and conferences, and provides a wide range of marketing services.

For additional information on any magazines or a complete catalog of ADVANSTAR Communications books, please write to ADVANSTAR Communications, Inc., 7500 Old Oak Boulevard, Cleveland, OH 44130.

Contents

Supplements   139

List of Symbols   167

Foreword

Gas chromatography is now more than 40 years old. In this period more than 100 general and specialized textbooks have been published on theory and practice, and the number of publications in this field is in the many tens of thousands. All of these are excellent texts and are very useful to the chromatographer who would like to enlarge his or her knowledge on various aspects of the technique.

There is, however, a problem that many practical chromatographers are facing: When they need only limited information, or when they are interested in just one relationship, such information is very difficult to find. Even the best subject index may fail in pointing to a specific subject. Also, many chromatographers simply do not have the time to go through hundreds of pages to find the information they need.

The purpose of this book is to fill this information gap. It is not a textbook of gas chromatography; it does not intend to explain the theory of GC or the function of the individual components of a gas chromatograph. Rather, its purpose is to help the practical chromatographer as a quick reference for finding the meaning of a term or the way it is calculated when reading a more complicated text, a journal publication, or doing day-to-day calculations.

The model for the original text of this book was a little handbook published in 1973 in Germany by Friedr. Vieweg & Sohn and entitled *Elementarmathematik griffbereit* (*Elementary Mathematics - At Hand*): this book did not try to give the theory of mathematics but rather, explained how to use it. This has been exactly our aim.

The first edition of this book was originally published in 1977 by the Perkin-Elmer Corporation, mainly as an aid to the participants of basic courses on gas chromatography. In the 15 years that have passed since then, over ten thousand copies of this small book have been printed. It was used not only by Perkin-Elmer but also for courses at quite different levels, from elementary courses to university classes.

This second, completely reorganized and enlarged edition is aimed at that readership: students in courses on gas chromatography at different levels, from elementary to advanced. In addition it should also serve as a handbook for anybody who wants to find information quickly on the meaning of a term, a relationship or the way some term can be calculated.

The title of this book indicates *gas chromatography* as its subject. However, probably 90 percent of the relationships discussed here can equally be related to liquid- or supercritical-fluid chromatography. In fact, where there are differences between the individual techniques, these are pointed out in the text.

Because of the book's general nature, we are not giving references to terms and relationships that have been in the public domain since the beginning of gas chromatography, and that have been discussed in many standardized texts. References are only given for newer terms that are less well known, or where further study might be needed. However, a special Supplement gives a selected list of textbooks dealing with various aspects of gas chromatography.

A special concern we considered was the proper use of terms and symbols. In this respect there is a lot of confusion in the literature: We even have cases in which author A uses symbols and names *a* and *b* to specify certain terms, while author B will discuss the same terms, but use the symbols in reversed order: *b* where author A used *a* and *a* where he used *b*. . . . Fortunately a new, completely revised *Nomenclature of Chromatography* has just been published by the International Union of Pure and Applied Chemistry (I.U.P.A.C.). This new nomenclature* has been followed in this book.

A peculiar contradiction existing in gas chromatography should be mentioned here: While most of the generalized terms used in GC are related to isothermal conditions, practical analysis is most likely carried out under programmed-temperature conditions. Therefore, we specially point out the terms that are also applicable for programmed-temperature conditions.

We hope that our colleagues will find this book useful in their day-to-day work.

May 15, 1993

Leslie S. Ettre
John V. Hinshaw

Pure Appl. Chem. **65,** 819–872 (1993).

PART I

Introduction

1.1 Principles of Chromatography

Chromatography is a physical method of separation in which the sample components to be separated (the *analytes*) are distributed between two phases, one of which is stationary (the *stationary phase*) while the other moves in a definite direction (the *mobile phase*).

This definition sets two basic criteria for a chromatographic process: that there are two phases present that are in contact with each other, and that one of these phases is stationary while the other is moving. Other separation methods also exist in practice that are similar to chromatography (e.g., field-flow fractionation or electrophoresis) but that do not meet these two criteria. These are therefore not chromatographic methods; however, they can be summarized under a broader term, *differential migration methods,* which also includes chromatography.

The chromatographic separation process occurs as a result of repeated sorption-desorption acts during the movement of the analytes. The separation is due to differences in the distribution of the individual sample components between the two phases. This distribution may be based on partitioning between the two phases (*partition chromatography*), or on adsorption-desorption processes (*adsorption chromatography*). The whole process can take place in a column containing the stationary phase through which the mobile phase is flowing (*column chromatography*), but the stationary phase may also occupy a planar surface (*planar chromatography*). The mobile phase may be a liquid (*liquid chromatography*, **LC**), a supercritical-fluid (*supercritical-fluid chromatography*, **SFC**) or a gas (*gas chromatography*, **GC**) and the stationary phase may be a solid, a gel or a liquid; if it is a liquid (*liquid phase*), it is distributed (*coated, immobilized* or *bonded*) on the surface of a solid that may or may not contribute to the separation process.

Depending on the physical state of the mobile phase, the characteristics of the stationary phase and the separation process, chromatography can be divided into a number of variants. Our present discussion concerns **gas chromatography**. However, most of the terms, definitions and relationships are also applicable to the other chromatographic techniques.

In gas chromatography the sample is volatized and then carried by the mobile phase (the *carrier gas*) into the column where the separation process takes place. At the end of the column, the analytes will emerge more or less separated in time. They are then detected and the detector signal is recorded. The detector signal is also used to establish the absolute and relative amounts of the analytes. If needed, the separated analytes can be directly conducted to another instrument, for example, a mass spectrometer or IR spectrophotometer, for further investigation (*hyphenated techniques*). During the whole process the system including the column is kept at a temperature sufficient to maintain the sample components in the vapor phase. The column may be under *isothermal conditions* or it is *temperature programmed* during analysis.

In summary, gas chromatography is a chromatographic technique capable of separating the individual components (analytes) contained in a mixture, in which the separation takes place in the vapor phase. By the proper monitoring of the column effluent, information is obtained that can be related to the identification of the individual sample components (*qualitative analysis*) as well as to the determination of their amounts present (*quantitative analysis*).

1.2 Components of a Gas Chromatograph

As shown in Figure 1, a gas chromatograph essentially consists of three integrated functions: sample introduction (evaporation), separation and detection. These functions are carried out by the sample introduction system, the column and the detector.

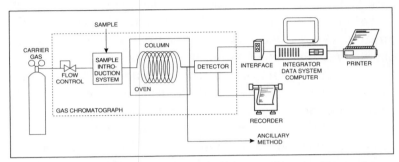

Figure 1. Basic components of a gas chromatographic system

The function of the *sample introduction system* (*injector*) is to vaporize the liquid samples and/or to permit the carrier gas to push the vapor in the form of a concentrated plug onto the beginning of the column with a minimum time lapse. Usually, the sample introduction system consists of a heated block (*flash vaporizer*) into which the liquid sample is injected with a *syringe*; however, sometimes the liquid sample is deposited directly at the front of the column (*on-column injection*), from where it is then evaporated during heating of the column. Gaseous samples are introduced into the carrier gas stream with the help of specially designed dosing systems (*gas sampling valve*).

The separation column contains the stationary phase which can be either an adsorbent or a liquid (*liquid phase*) distributed in the form of a thin film on the surface of a solid *support*. The solid support may consist of porous particles packed into the column tube (*packed columns*), but the liquid stationary phase may also be distributed on the inner wall of the column tubing (*open-tubular* or *capillary columns*). In the latter case the tube wall may be smooth, without having its physical character changed (*wall-coated open-tubular* [**WCOT**] *columns*) or it may be modified by building up or depositing a porous layer on the original tube wall which in turn is coated with the liquid phase (*porous-layer open-tubular* [**PLOT**] *columns*). This porous layer may also consist of support particles deposited on the tube wall (*support-coated open-tubular* [**SCOT**] *columns*).

The separation column is placed in a *temperature-controlled oven*. This is either maintained at a constant temperature during analysis (*isothermal operation*) or its temperature is increased with time in a manner that is reproducible from analysis to analysis (*programmed-temperature operation*).

The individual components separated in the column emerge with the carrier gas and are detected by continuously monitoring some physical or chemical properties of the effluent. A number of *detectors* are available; among these there are two — the thermal-conductivity and the flame-ionization detectors — that are most universally used. The *thermal-conductivity* (or *hot-wire*) *detector* (**TCD, HWD**) measures the differences in the thermal conductivity of the column effluent vs. that of the pure carrier gas, while the *flame-ionization detector* (**FID**) breaks down the sample components into ions in a hydrogen flame, which are collected to form a useful signal. In addition to these two general-purpose detectors that respond to most substances present in the sample, a number of other detectors also exist that are selective to certain types of substances or functional groups.

Besides these three main functions, two others are also important for successful gas chromatographic separation. These are related to the carrier gas and the detector readout.

Separation in the column takes place in a continuously moving inert *carrier gas* stream. Thus, one needs a source for the carrier gas, and its flow and/or pressure must be controlled permitting the reproduction of its conditions (flow rate, inlet pressure). The most universally used carrier gases are helium, hydrogen and nitrogen.

The detector signal is displayed on a strip-chart *recorder*, or the read-out of the *data system*, producing the familiar chromatograms consisting of individual peaks. In modern gas chromatography, data handling and interpretation may be carried out with the help of more-or-less sophisticated data systems.

The chromatogram provides two types of data used for *qualitative* and *quantitative* interpretation. Due to the time-ordered elution of the components from the column, their *retention time* — the time that passed between sample introduction and their emergence from the column — is used for component identification. Furthermore, since the change in the physical or chemical properties of the column effluent during the elution of a separated sample component is proportional to the amount present, the detector's response is used, after appropriate calibration, for the establishment of the absolute and relative amounts (concentrations) of the sample components.

1.3 The Chromatographic Separation Process

This generalized discussion refers to column chromatography; however, essentially the same principles exist also in planar chromatography.

Let us assume that we introduce a two-component mixture at the front of the chromatographic column. One of the two components was selected so that it will not be retained at all by the stationary phase but will always stay in the mobile phase. Thus, it will pass through the column with the speed of the mobile phase. The second component, however, will go through repeated sorption-desorption steps during its passage along the stationary phase and, therefore, will emerge later than the first component.

The separation of such a binary mixture is shown graphically in Figure 2 on page 6. The open circles represent the molecules of that sample component that is completely insoluble in the liquid phase, while the black dots represent the molecules of a sample component that is soluble in it. For this illustration we assumed that upon equilibrium, equal numbers of molecules will be in both the mobile phase and the stationary phase. The column is divided into small segments — "theoretical plates" — and for the illustration, it is assumed that the passage through the column is composed of consecutive travel and equilibrium

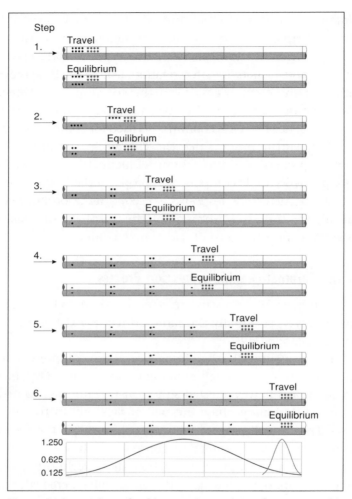

Figure 2. Separation of a binary mixture in a chromatographic column. Top part: *Mobile (gas) phase;* bottom part: *Stationary (liquid) phase.* Open circles *represent the molecules of a component insoluble in the stationary phase, while the* black dots *represent the molecules of a sample component which is soluble in it. The* vertical lines *divide the column into "theoretical plates."*

steps. In the "travel" step, the mobile phase of one plate (and hence, the sample component molecules present in it) moves to the next plate, while in the equilibrium step, the molecules present in a plate will be distributed between the two phases according to their distribution ratio (in this case, 1:1). Figure 2 illustrates six consecutive travel and equilibrium steps: It can be seen that already in such a short length of the column the two sample components have separated.

A chromatographic column consists of many "theoretical plates." Thus, during the chromatographic separation process the sample components are more and more separated and will elute from the column at different times. The first component to elute will be the one that is not retained at all by the stationary phase. The other sample components elute in sequence according to how strongly they were retarded by the stationary phase. The more a component is dissolved by the stationary phase, the later it will elute from the column.

Since each sample component will be present in the column in a number of "theoretical plates" (see the black dots in Figure 2), its concentration in these "plates" will be different, resulting in a Gaussian distribution which can already be seen in Figure 2. The elution time corresponding to the maximum of this distribution curve — the chromatographic *peak* — represents the *total retention time* of the particular analyte.

The sorption-desorption steps result in a concentration distribution at equilibrium between the two phases. The ratio of the concentration of the analyte in the stationary phase to its concentration in the mobile phase is the *distribution constant* or *partition coefficient* (*K*) of the analyte:

$$K = \frac{concentration\ in\ stationary\ phase}{concentration\ in\ mobile\ phase} \qquad \text{eq.1.1}$$

The partition coefficient expresses how strongly a substance is retained in the stationary phase relative to the mobile phase. Sample components with smaller partition coefficients elute before those with larger partition coefficients.

In partition chromatography, concentration is expressed as amount per volume of the phase:

$$K = \frac{W_{i(S)}/V_S}{W_{i(M)}/V_M} = \frac{W_{i(S)}}{W_{i(M)}} \cdot \frac{V_M}{V_S} \qquad \text{eq.1.2}$$

where $W_{i(S)}$ and $W_{i(M)}$ are the amounts of the analyte in the stationary and mobile phases, while V_S and V_M are the volumes of the stationary and mobile phases, respectively. There are, however, also other ways to express concentration; for example, in adsorption chromatography, it may be expressed per unit surface area, etc.

The partition coefficient (distribution constant) is a substance-specific parameter which, in a given stationary phase and at a given temperature, is independent of other operating parameters such as the amount of the two phases in the column. In general, we also assume that the partition coefficient is independent of the total concentration of the analyte.

Thermodynamically, the partition coefficient can be expressed as a function of the saturated vapor pressure (p^o) and activity coefficient (γ^o) of the analyte, the density (ρ_S) and molecular mass (M_S) of the stationary phase and the (absolute) column temperature (T_c):

$$K = \frac{\rho_S \cdot R \cdot T_c}{p^o \cdot \gamma^o \cdot M_S} \qquad \text{eq.1.3}$$

where R is the gas constant.

The temperature dependency of the partition coefficient can be expressed by the following equation:

$$\log K = \frac{A}{T_c} + B \qquad \text{eq.1.4}$$

where T_c is the absolute column temperature and A and B are constants. In other words, at higher temperatures the value of K will be smaller and the analyte will elute sooner.

In a relatively narrow temperature range the $\log K$ vs. T_c relationship can be approximated by a linear equation:

$$\log K \approx A' \cdot T_c + B' \qquad \text{eq.1.5}$$

where A' and B' are again constants. The value of A' will be negative.

From the foregoing discussion it can be seen that the chromatographic separation process is controlled by the selective affinity of sample components for the stationary and mobile phases as well as by the temperature.

PART II

Data Measured from the Chromatogram

If a detector is attached to the end of the chromatographic column to monitor the presence of the sample components in the mobile (gas) phase eluting from the column, the recorder will display a chromatogram similar to the one shown in Figure 3 on page 10. There are a number of important values which can be established directly from the chromatogram.

2.1 Retention Times

As mentioned in the previous part, the first component to elute is the one that is not retained at all by the stationary phase. The time measured between the instant of sample introduction and the maximum of this peak (t_M) is called the *retention time of the unretained solute,* or, in gas chromatography, the *gas holdup time.* The term *air peak time* is also used often since in most stationary phases air represents a typical unretained solute. When a flame-ionization detector is used, methane is usually applied as an analyte giving the gas holdup time. In liquid chromatography the situation with the holdup time is not as simple as in gas chromatography; for details consult general textbooks on LC.

The time between the instant of sample introduction and

the maximum of the retained peak is the *total retention time* (t_R) of this analyte. This value is the sum of two time values:

$$t_R = t_M + t'_R \qquad\qquad \text{eq.2.1}$$

The first, t_M, is the holdup time as defined above, while t_R' represents the time the molecules of the analyte spend in the stationary phase. We call this the *adjusted retention time*.

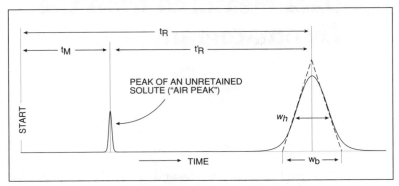

Figure 3. A typical chromatogram

Two notes are in order here.

(1) The adjusted retention time represents correction of the total retention time for the holdup in the mobile phase. Another correction used in gas chromatography considers the compressibility of the gas in the column. This correction is used for the various *retention volumes* representing not the time elapsed between sample introduction and peak maximum, but the mobile phase volume flowing through the system during this time. From the various retention volumes additional retention time expressions such as the *corrected retention time* ($t_R°$) and the *net retention time* (t_N) may be derived. We shall define them in Part VI.

(2) When reading the literature one must be careful with the names assigned to t_R': Sometimes this is called the "retention time," and one may also find it as the "corrected retention time" or even the "net retention time," although these two terms have different mean-

ings. This error stems from two sources. The first involves the difficulty in translating from German, where there is not such a clear-cut difference in the literature between these three adjectives ("adjusted," "corrected" and "net"), and where the adjusted retention time (volume) has also been called the *Nettoretentionszeit* (*Nettoretentionsvolumen*). The second reason for the mix-up comes from liquid chromatography. Because there the compression correction factor (j) used in the calculation of both the corrected and net retention times (volumes) is equal to unity, the net retention time (volume) and the adjusted retention time (volume) will have the same value.

Both the holdup time and the retention time can be established by measuring on the recorder chart paper the distance between the start point and the maximum of the respective peak. This distance is then converted into time units by the application of the speed of the recorder chart paper:

$$retention\ time\ = \frac{distance\ on\ the\ chart\ paper}{chart\ paper\ speed} \qquad eq.2.2$$

Obviously the same dimensions must be used in both terms present in the fraction on the right-hand side of eq.2.2. Modern data systems measure the retention time and display its value directly.

2.2 Determination of the Gas Holdup Time

From a chromatogram we usually measure t_R. On the other hand, for practically all chromatographic calculations (except the plate number) we need t_R'. Therefore, it is important to establish the gas holdup time. This can be done by direct measurement using air or, in the case of a flame-ionization detector, methane.*

Another possibility is to calculate the gas holdup time. This can be done if the retention times of three members of a ho-

---

* Some columns, such as ones containing molecular sieves, retain and separate the components of air, as well as methane.

mologous series (measured under identical analytical conditions) are known. This calculation is based on the semilog relationship between the adjusted retention time and the number of carbon atoms (c_n) in the molecule of the members of a homologous series:

$$\log t'_R = a \cdot c_n + b \qquad \text{eq.2.3}$$

There are two criteria for this calculation: The homologs should have a somewhat higher carbon number (if possible, above C_5), and their carbon number difference must be constant. The first criterion is important because the semilog relationship between the adjusted retention time and the carbon number for a homologous series is generally not linear for the first members of a homologous series. Constant carbon number difference means that if c_n represents the carbon number and subindices 1, 2 and 3 refer to the three homologs with increasing retention time, then:

$$c_{n2} - c_{n1} = c_{n3} - c_{n2} \qquad \text{eq.2.4}$$

The calculation is fairly straightforward and is based on the following relationship (for details see Supplement No.I):

$$t_M = t_{R2} - \frac{x_1 \cdot x_3}{x_3 - x_1} \qquad \text{eq.2.5}$$

where

$$x_1 = t_{R2} - t_{R1} \qquad \text{eq.2.6a}$$

$$x_3 = t_{R3} - t_{R2} \qquad \text{eq.2.6b}$$

2.3 Peak Widths

Ideally the chromatographic peak has the shape of a Gaussian distribution curve. The area and height of this peak are a function of the amount of solute present, while its width is a function of band spreading in the column.

In the case of a Gaussian distribution curve the following relationship is valid:

$$m = m_{max} \cdot e^{-(x^2/2\sigma^2)} \qquad\qquad \text{eq.2.7}$$

where m is the peak height at any point, m_{max} is the peak height at maximum, σ is the standard deviation of the peak and x is the distance from the ordinate (the perpendicular to the base line at peak maximum); in other words, x is half of the peak width (w) at that point. It is convenient to introduce the function ϕ:

$$\phi = m / m_{max} \qquad\qquad \text{eq.2.8}$$

which is the fraction of the peak height at a given point relative to the peak height at maximum (which are the values on the ordinate in Figure 4, page 14). Introducing w and ϕ into eq.2.7, we can write it in the following form:

$$\phi = e^{-(w^2/8\sigma^2)} \qquad\qquad \text{eq.2.9}$$

From this equation we can express the peak width at any point as a function of ϕ and σ:

$$w = 2\sigma\sqrt{2(-ln\phi)} = 2\sigma\sqrt{2ln(1/\phi)} \qquad\qquad \text{eq.2.10}$$

Figure 4 indicates the peak widths at various peak heights as a function of the standard deviation of the peak. From these, three are important:

- the peak width at half height (w_h): This is the peak width measured at 50% of the maximum peak height;
- the peak width at base (w_b): This is the segment of the base line cut out by the two tangents drawn to the inflection points of the Gaussian curve;
- the peak width at inflection points (w_i): This is the peak width measured between the inflection points.

Their values are:

$$w_h = \left(2\sqrt{2ln2}\right)\sigma = 2.354\sigma \qquad\qquad \text{eq.2.11a}$$

$$w_b = 4\sigma \qquad\qquad \text{eq.2.11b}$$

$$w_i = 2\sigma \qquad\qquad \text{eq.2.11c}$$

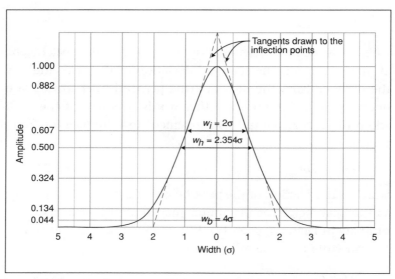

Figure 4. Characteristic widths of a Gaussian peak: w_i = *peak width at the inflection points;* w_h = *peak width at half height;* w_b = *peak width at base.*

From eqs.2.11a and 2.11b we can also derive the relationship between w_h and w_b:

$$w_b = \frac{4}{2\sqrt{2 ln2}} w_h = 1.699 w_h \qquad \text{eq.2.12}$$

In the case of a standard strip-chart recorder, one is measuring the peak widths, for example, with the help of a magnifying glass with an etched scale, usually divided to 0.1 mm. In modern data systems, the peak width is printed out in a report: the system calculates it from the integrated peak area (see Section 2.4), or by measuring directly from the peak shape.

2.4 Peak Area

The area of a Gaussian peak (A) can be expressed as a function of the peak height at maximum (m_{max}) and the standard deviation (σ) of the peak:

$$A = m_{max} \cdot \sigma\sqrt{2\pi} \qquad \text{eq.2.13}$$

If the peak area has to be established manually from the chromatogram, this is usually done by multiplying the peak height at maximum (m_{max}) with the peak width at half height (w_h), assuming that the peak is Gaussian in shape:

$$A' = m_{max} \cdot w_h \qquad \text{eq.2.14}$$

The value of A' calculated in this way is not equal to A, the true integrated peak area, but is 94% of its value. This can be easily derived from eq.2.13, which can be written in the following way:

$$m_{max} \cdot \sigma = A \frac{1}{\sqrt{2\pi}} \qquad \text{eq.2.15}$$

Since $w_h = 2.354\sigma$ (cf. eq.2.11a), we can write that:

$$\frac{m_{max} \cdot w_h}{2.354} = \frac{A}{\sqrt{2\pi}}$$

$$m_{max} \cdot w_h = A \frac{2.354}{\sqrt{2\pi}} = 0.94A \qquad \text{eq.2.16}$$

2.5 Peak Symmetry

In an ideal case, a peak is symmetrical. This means that both its front and back have the same slope or — expressed differently — the frontal part of the peak width at base (a) has an identical length to the back part (b) of the peak width (see Figure 5, A, page 16):

$$a = b \qquad \text{eq.2.17}$$

If the peak *tails,* this means that its front is steeper than its rear; in other words:

$$b > a \qquad \text{eq.2.18}$$

The opposite is also possible; in this case the peak is *fronting*:

$$a > b \qquad \text{eq.2.19}$$

There are various ways to mathematically express peak asymmetry. In the most generally accepted method the front

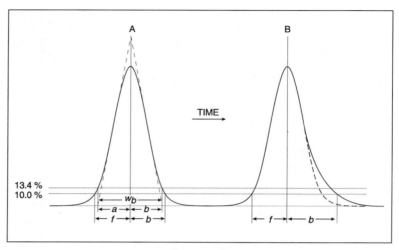

Figure 5. Peak symmetry calculation

(f) and back (b) part of the peak width at 10% of the maximum peak height is measured (see Figure 5, B). Next the value of their absolute difference (Δw) is calculated:

$$\Delta w = |b - f| \qquad \text{eq.2.20}$$

There is a practical justification for measuring the width at 10% of the peak height. As seen in Figure 4 on page 14, the actual peak width measured at 13.4% of the peak height is equal to the peak width at base ($w_b = 4\sigma$). It is certainly easier to measure the actual peak width than to draw tangents to the inflection points and then establish the length of the base line cut out by these two tangents. The difference between the actual peak width at 13.4% or at 10% of the peak height is relatively small; with the help of eq.2.10 one can calculate that the peak width at 10% peak height is 4.3σ instead of 4σ, i.e., the difference is 7.5%.

Knowing Δw, the peak symmetry (As) is calculated as:

$$As = \frac{w}{w - \Delta w} \qquad \text{eq.2.21}$$

where w is the actual peak width at 10% of the peak height:

$$w = f + b \qquad \text{eq.2.22}$$

If $b = f$, $\Delta w = 0$, and the peak is symmetrical ($As = 1.00$).

Since in eq.2.20 we take the absolute difference without regard to the sign, the numerical value of As will not indicate whether the particular case is tailing or fronting: This must be explicitly stated.

Let us examine the meaning of eq.2.21, first considering a tailing peak. In such a case the front part of the peak is still fairly Gaussian, particularly if Δw is relatively small. This means that one can assume that a Gaussian peak at the same position would have the same front part as the present peak, while its back part would be a mirror image of the frontal part (see the broken line in Figure 5, B, page 16). The width of this peak would be *2f*, which is equal to $(w - \Delta w)$. In other words, the denominator of the fraction in eq.2.21 gives the width of the corresponding symmetrical peak. Therefore, the asymmetry term is equal to:

$$As = \frac{actual\ peak\ width}{width\ of\ symmetrical\ peak} \qquad \text{eq.2.23}$$

The same is true for fronting peaks with the difference that now the backside of the peak is equivalent to the corresponding Gaussian peak.

2.6 Peak Volume

This expression is often used in calculations related to the detectors. The peak volume (V_P) is equal to the mobile phase volume corresponding to the peak width measured from the chromatogram. Usually, the peak width at half height (w_h) is used for this purpose:

$$V_P = w_h \cdot F_c \qquad \text{eq.2.24}$$

where F_c is the mobile phase flow rate measured at column outlet at ambient pressure and corrected to column temperature (see Section 5.1).

PART III

Reduced Terms Expressing Retardation in the Column

As seen earlier, the *total retention time* (t_R) represents the time the analyte is retarded by the column. This period can be divided into two segments: the *adjusted retention time* (t_R'), representing the time the analyte is retarded by the stationary phase, and the *holdup time* (t_M), representing the time it spends in the mobile phase, traveling through the column:

$$t_R = t_M + t_R' \qquad \text{eq.2.1}$$

The absolute values of these retention times depend on the flow rate and column length; therefore, their relative (reduced) values are more descriptive of the chromatographic separation process.

There are five such dimensionless reduced terms expressing relative retardation by the two phases in the column. While the retention factor (k) is the most prevalently used term, the others mentioned in this section are found throughout the fundamental chromatographic relationships for retention, separation and detection. Table I on page 19 presents numerical values of these terms.

k	k+1	(k+1)/k	k/(k+1)	1/(k+1)
0.1	1.1	11.0000	0.0909	0.9091
0.5	1.5	3.0000	0.7500	0.6667
1.0	2.0	2.0000	0.5000	0.5000
1.5	2.5	1.6667	0.6000	0.4000
2.0	3.0	1.5000	0.6667	0.3333
2.5	3.5	1.4000	0.7143	0.2857
5.0	6.0	1.2000	0.8333	0.1667
10.0	11.0	1.1000	0.9091	0.0909
15.0	16.0	1.0667	0.9375	0.0625
20.0	21.0	1.0500	0.9524	0.0476
25.0	26.0	1.0400	0.9615	0.0385
50.0	51.0	1.0200	0.9804	0.0196

Table I. Values of certain reduced terms, as a function of k

3.1 Retention Factor or Capacity Ratio

This term, the symbol* for which is k or k', is a measure of the time the analyte resides in the stationary phase relative to the time it spends in the mobile phase. In other words, it expresses how much longer an analyte is retarded by the stationary phase than it would take to travel through the column with the velocity of the carrier gas:

$$k = t'_R/t_M \qquad \text{eq.3.1}$$

If the distribution constant (the partition coefficient) is independent of analyte concentration, then the retention factor is also equal to the ratio of the amounts of the analyte in the stationary ($W_{i(S)}$) and mobile phases ($W_{i(M)}$) respectively, at equilibrium:

$$k = \frac{W_{i(S)}}{W_{i(M)}} \qquad \text{eq.3.2}$$

---

*Particularly in liquid chromatography, the symbol k' is often used for this term.

In the past this term has been indicated by different names such as the partition ratio, capacity ratio, capacity factor, mass distribution ratio or the retention factor. The newest I.U.P.A.C. nomenclature suggests the use of the name *retention factor* for this term.

As we shall see later in Section 4.7, in a given column the retention factor is directly proportional to the distribution constant (partition coefficient). Thus, the temperature dependency of k can be described similarly to eq.1.4:

$$\log k = \frac{a}{T_c} + b \qquad\qquad \text{eq.3.3a}$$

But, in a limited temperature range, it can be approximated as:

$$\log k \simeq a'T_c + b' \qquad\qquad \text{eq.3.3b}$$

where T_c is the absolute temperature of the column and a, b, a' and b' are constants. The value of a' in eq.3.3b is negative: the higher the temperature the smaller the retention factor. According to general GC practice a temperature difference of about 15-25°C results in a change in k by a factor of about two.

3.2 Total Retention Relative to Holdup Time

This term is a measure of the total time the analyte resides in the column relative to the time it spends in the mobile phase. In other words, it expresses how much longer the analyte is retarded by the column than it would take to travel through it with the velocity of the mobile phase:

$$\frac{t_R}{t_M} = \frac{t'_R + t_M}{t_M} = k + 1 \qquad\qquad \text{eq.3.4}$$

The familiar expression of the total retention time follows from this relationship:

$$t_R = t_M (1+k) \qquad\qquad \text{eq.3.5}$$

3.3 Total Retention in the Column Relative to Retardation by the Stationary Phase

This is a measure of the total time the analyte resides in the column relative to the time it resides in the stationary phase. In other words, this term expresses how much longer the analyte is retarded by the column than it is retarded by the stationary phase only:

$$\frac{t_R}{t_R'} = \frac{t_R' + t_M}{t_R'} = \frac{k+1}{k} \qquad \text{eq.3.6}$$

3.4 Retardation by the Stationary Phase Relative to Total Retention in the Column

This term is a measure of the time the analyte resides in the stationary phase relative to the total time it resides in the column. In other words, it expresses how much longer an analyte is retarded by the stationary phase than the total time it is retarded by the column:

$$\frac{t_R'}{t_R} = \frac{t_R'}{t_R' + t_M} = \frac{k}{k+1} \qquad \text{eq.3.7}$$

3.5 Retardation Factor

This term is a measure of the time the analyte resides in the mobile phase relative to the total time it resides in the column. In other words, it expresses the fraction of time the analyte spends in the mobile phase relative to the total time it is retarded by the column:

$$R = \frac{t_M}{t_R} = \frac{t_M}{t_R' + t_M} = \frac{1}{k+1} \qquad \text{eq.3.8}$$

From eq.3.8 we can express that:

$$k = \frac{1-R}{R} \qquad \text{eq.3.9}$$

PART IV

The Chromatographic Column

The chromatographic column tube is defined by its *length* (*L*), *inner diameter* (*d_c*) and *inner radius* (*r_c*). The column may be **packed**; in this case the particles of the solid stationary phase or of the support (which is coated by the stationary phase) are characterized by the *particle diameter* (*d_p*). In the case of **wall-coated open-tubular columns**, the inner wall of the column tube is coated with a relatively thin stationary (liquid) phase film having a film thickness of d_f. In the case of **porous-layer open-tubular columns** the situation is more complicated. There is a porous layer on the inner wall of the tube, the surface of which is larger than the geometric surface of the tube, and this surface is coated by the (liquid) stationary phase. It is also possible that the porous layer consists of an adsorbent or a solid stationary phase.

There are a number of terms related to the chromatographic column.

4.1 Column Tube Volume and Inner Surface Area

The column tube's inner volume (V_c) is expressed as:

$$V_c = d_c^2 \, \pi \cdot L \, / \, 4 = r_c^2 \, \pi \cdot L \qquad \text{eq.4.1}$$

The geometric inner surface area (A_{geom}) of the column tube is:

$$A_{geom} = d_c \pi \cdot L = 2 r_c \pi \cdot L \qquad \text{eq.4.2}$$

4.2 Cross-Sectional Area of the Column

The cross-sectional area (A_c) of the column tube is:

$$A_c = d_c^2 \pi / 4 = r_c^2 \pi \qquad \text{eq.4.3a}$$

In the case of **wall-coated open-tubular columns**, this full area is available to the mobile phase flow. In this respect we usually neglect the very thin stationary phase coating. In the case of thick-film open-tubular columns, the proper correction may be made:

$$A_c = (r_c - d_f)^2 \pi \qquad \text{eq.4.3b}$$

It should be noted that in the case of a 0.53-mm ID open-tubular column with a 5-µm liquid phase film, an error of 3.9% is made when using eq.4.3a instead of eq.4.3b.

In the case of **packed columns** (and also of **porous-layer open-tubular columns**), only a fraction of the geometric cross-sectional area is available to the mobile phase flow, the rest being occupied by the solid particles. The fraction of the geometric cross-sectional area available to the mobile phase flow is expressed by the *interparticle porosity* (ε). This is equal to the ratio of the actual volume occupied by the mobile phase between the solid particles (V_o: the *interparticle volume of the column*) and the geometric volume of the column (V_c: see eq.4.1):

$$\varepsilon = V_o / V_c \qquad \text{eq.4.4}$$

In the case of regular packed columns containing glass beads or diatomaceous earth type support particles, the value of ε is close to 0.40. For a (wall-coated) open-tubular column, $\varepsilon = 1.00$ because the whole cross-section is available to the mobile phase flow.

As a conclusion of this, the active cross-sectional area of a packed column is:

$$A_c = \varepsilon (r_c^2 \pi) \qquad \text{eq.4.5}$$

4.3 Mobile Phase Volume of the Column

The volume of the mobile phase in the column (V_M) is for a **packed column**:

$$V_M = \varepsilon(r_c^2\pi)L \qquad \text{eq.4.6}$$

and for a (wall-coated) **open-tubular column**:

$$V_M = (r_c^2\pi)L \qquad \text{eq.4.7a}$$

or

$$V_M = (r_c - d_f)^2\,\pi \cdot L \qquad \text{eq.4.7b}$$

In **liquid chromatography** V_M is equal to the holdup volume, established from the mobile phase holdup time (t_M) and the mobile phase flow rate (F_c) (see Section 6.1), and this is also equal to the volume of the mobile phase in the column. In **gas chromatography**, however, we must distinguish between two values because of the compressibility of the carrier gas:

- here V_M is the *gas holdup volume* calculated from the gas holdup time (t_M) and the carrier gas flow rate (F_c):

$$V_M = t_M \cdot F_c \qquad \text{eq.4.8a}$$

- on the other hand, the *volume of the carrier gas in the column* (called V_G) is calculated from V_M by utilizing the compression correction factor (j) (see Section 5.3):

$$V_G = V_M \cdot j = V_M^o \qquad \text{eq.4.8b}$$

V_M° is called the *corrected gas holdup volume* (see Section 6.2).

4.4 Volume of the Stationary (Liquid) Phase in the Column

In the case of **packed columns**, the volume of the stationary phase cannot be established directly. It can, however, be calculated by utilizing eq.4.21 which relates the distribution constant (partition coefficient: K) to the retention factor (k) and the phase ratio (β):

$$K = k \cdot \beta = k(V_M/V_S) \qquad \text{eq.4.9a}$$

From this relationship we can write that:

$$V_S = V_M (k / K)$$ eq.4.9b

where V_S and V_M are the volumes of the stationary and mobile phases, respectively. Thus, if we know V_M, k and K, then V_S can be calculated. The value of k can be measured directly from the chromatogram while $V_M (V_G)$ can be calculated with help of eqs. 4.6, 4.7a-b or 4.8b. The value of K can be established indirectly by carrying out a parallel measurement on a column (e.g., an open-tubular column) coated with the same stationary phase and operated at the same temperature for which we know the value of the phase ratio (calculated from the tube radius and stationary phase film thickness: eq.4.19). The partition coefficient of the analyte on this column will be identical to that on the packed column of interest. Thus, if we measure the retention factor k of the same analyte on this column, the partition coefficient can be calculated as:

$$K = k \cdot \beta$$

In the case of (wall-coated) **open-tubular columns**, the volume of the liquid phase (V_L) can be calculated from the film thickness (d_f) (see Section 4.5) and the (geometric) surface area of the column tube:

$$V_L = (2 r_c \pi) L \cdot d_f$$ eq.4.10

4.5 Liquid Phase Film Thickness

In the case of **packed columns**, the average thickness of the coating on the support particles can be approximated if one knows the total volume of the liquid phase (V_L), the total amount of the support in the column (W_{sup}) and the specific surface area of the support (S_{sup}, m²/g). From the latter two one can calculate the total coated surface area (A_{coated}):

$$A_{coated} = W_{sup} \cdot S_{sup}$$ eq.4.11

and then, from this and the volume of the liquid phase, one can calculate the average thickness of the liquid phase coating (d_f):

$$d_f = V_L / A_{coated}$$ eq.4.12

This is, however, only an approximate value since it is well known that coating in packed columns is usually not uniform; e.g., there are "puddles" where particles touch each other.

In the case of **wall-coated open-tubular columns** the thickness of the liquid phase coating (d_f) can be established in two ways:

- by comparing the column at identical temperature with another column coated with the same stationary (liquid) phase the film thickness of which is known;

- by calculating from static coating data.*

In the first case, the calculation is based on the fact that the partition coefficient (distribution constant: K) of an analyte at a given temperature is independent of the actual column used (see Section 1.3). Since for open-tubular columns (see Section 4.8):

$$K = k \cdot \beta = \frac{k \cdot r_c}{2d_f}$$ eq.4.13

where k is the retention factor of the analyte, β is the phase ratio of the column (cf. Section 4.8, below), and r_c is the inner radius of the column tube, one can write for two columns coated with the same liquid phase and operated at the same temperature, that:

$$d_{f2} = d_{f1} \frac{k_2}{k_1} \cdot \frac{r_{c2}}{r_{c1}}$$ eq.4.14

In the second case the value of the film thickness (d_f) can be established from the composition of the coating solution:

$$d_f = \frac{r_c}{2} \cdot \frac{c_L \cdot \rho}{100 - (c_L / \rho)}$$ eq.4.15

where r_c is the column tube's inner radius, c_L is the concentration of the liquid phase in the coating solution (in wt/vol %) and ρ is the density of the liquid phase (in g/mL). The derivation of this equation is given in Supplement No. II.

In the case of **porous wall-coated open-tubular columns**, the average thickness of the liquid phase coating can only be

---

* There are also some empirical relationships that permit the approximate calculation of the film thickness from dynamic coating data.

established when knowing the actual coated surface area in the column and the volume of the liquid phase in the column (cf. eq.4.12).

4.6 Reduced Liquid Phase Film Thickness

This dimensionless value was introduced by Schoenmakers [1]:

$$\delta_f = \frac{d_f}{d_c}\sqrt{\frac{D_M}{D_S}} \qquad\qquad \text{eq.4.16}$$

where d_f is the film thickness, d_c is the inner diameter of the column tube, and D_M and D_S are the diffusion coefficients of the analyte in the mobile and stationary phases, respectively. In Section 11.6 we shall discuss some utilization of the reduced liquid phase film thickness.

4.7 Liquid Phase Loading

In gas chromatography with packed columns, this term characterizes the relative amount of the liquid phase in the column packing. It is equal to the weight percent ($W_L\%$) of the liquid (stationary) phase in the total packing:

$$W_L\% = \frac{100\cdot W_L}{W_{Sup} + W_L} \qquad\qquad \text{eq.4.17}$$

where W_L and W_{Sup} are the weight of the liquid phase and the support in the column, respectively.

4.8 Phase Ratio

The phase ratio (β) of a column is the ratio of the volumes of the mobile (V_M) and stationary (V_S) phases in the column:

$$\beta = V_M / V_S \qquad\qquad \text{eq.4.18a}$$

In gas chromatography one must substitute the volume of the gas phase in the column ($V_G = V_M^\circ$) for V_M, while we may use the symbol V_L (volume of the liquid phase) for V_S:

$$\beta = V_M^\circ / V_L = V_G / V_L \qquad\qquad \text{eq.4.18b}$$

As indicated above, V_G is not equal to the gas holdup volume calculated from the gas holdup time and the carrier gas flow rate but to the *corrected* gas holdup volume ($V_M{}°$) in which gas compression was considered (see Section 4.3).

In the case of **wall-coated open-tubular columns** both V_G and V_L can be established from the column dimensions and the liquid phase film thickness (cf. eqs.4.7a–b and 4.10). For thin-film columns where $r_c \gg d_f$, the phase ratio can be written as:

$$\beta = \frac{(r_c^2 \pi) L}{(2 r_c \pi) L \cdot d_f} = \frac{r_c}{2 d_f}$$

eq.4.19

If a thicker liquid phase coating is present, we may use eq.4.7b instead of eq.4.7a:

$$\beta = \frac{[(r_c - d_f)^2 \pi] L}{(2 r_c \pi) L \cdot d_f} \approx \frac{r_c - 2 d_f}{2 d_f}$$

eq.4.20

For example, in the case of a column with 0.53 mm I.D. and a film thickness of 5 μm, the use of eq.4.19 instead of eq.4.20 results in a 3.9% difference in the value of the phase ratio.

Considering eqs.1.2 and 3.2, we can relate the distribution constant (partition coefficient) (K) to the retention factor (k) and the phase ratio:

$$K = k \cdot \beta$$

eq.4.21

This relationship will be discussed in detail in Section 11.1.

4.9 Particle Size

In GC packed columns the diameter of the support or solid stationary phase particles is much larger than that of those used in modern liquid chromatography. In LC the particle size is usually given in micrometers; on the other hand, in GC it is usually characterized by the mesh or screen numbers. Two sieve series exist in the U.S.A.: the U.S. Standard Sieve Scale and the Tyler Sieve Series.

Screens of the U.S. Standard Sieves are designated by arbitrary numbers. The system is based on the 18-mesh screen in which the aperture is 1.00 mm (0.0394 in.). The aperture ratio of two consecutive screens is the fourth root of two.

The Tyler Sieve Series is characterized by the number of meshes per inch. The original system was based on the 200-mesh screen, and the ratio of apertures of two consecutive screens was the square root of two. Since this series gave a too-wide particle range between two screens, intermediate screens (Double Tyler Series) were added: The aperture ratio of two consecutive screens in the complete series is the same as in the U.S. Standard Sieves (fourth root of two).

Tables II and III list the specifications of the two sieve series. To establish the actual particle size range of a mesh range, one takes the aperture values corresponding to the top and bottom screen. For example, if the mesh range is specified as 80/100 (U.S. Standard), the aperture of the top screen (80 mesh) is 0.177

Sieve no. ("mesh")	Aperture	
	in.	mm
30	0.0232	0.59
35	0.0197	0.50
40	0.0165	0.42
45	0.0138	0.35
50	0.0117	0.297
60	0.0098	0.250
70	0.0083	0.210
80	0.0070	0.177
100	0.0059	0.149
120	0.0049	0.125
140	0.0041	0.105
170	0.0035	0.088
200	0.0029	0.074
230	0.0024	0.062
270	0.0021	0.053
325	0.0017	0.044

*Table II. Sieve specifications: U.S. Standard Sieves ***

* Source: ref. [2]

Tyler Series mesh	Double Tyler Series mesh	Aperture	
		in.	mm
28		0.0232	0.589
	32	0.0195	0.495
35		0.0164	0.417
	42	0.0138	0.351
48		0.0116	0.295
	60	0.0097	0.246
65		0.0082	0.208
	80	0.0069	0.175
100		0.0058	0.147
	115	0.0049	0.124
150		0.0041	0.104
	170	0.0035	0.088
200		0.0029	0.074
	250	0.0024	0.061
270		0.0021	0.053
	325	0.0017	0.043
400		0.0015	0.037

*Table III. Sieve specifications: Tyler Sieve Series *

mm and of the bottom screen (100 mesh) 0.149 mm. The same mesh range in the Tyler Screen Series corresponds to 0.175–0.147 mm particle diameter range.

4.10 Specific Permeability

A column through which a fluid is flowing offers a certain resistance to the flow. A convenient parameter for defining this resistance is the *specific permeability* (B_o) of the column.

In the case of **packed columns**, the specific permeability is proportional to the square of the particle diameter (d_p) and to

---

* Source: ref. [2]

a term including the interparticle porosity (ε) of the column (see Section 4.2):

$$B_o = \frac{d_p^2}{180} \cdot \frac{\varepsilon^3}{(1-\varepsilon)^2} \qquad \text{eq.4.22a}$$

This, the so-called Kozeny-Carman equation, can also be written by including the mobile phase inlet (p_i) and outlet (p_o) pressures, the mobile phase velocity at column outlet (u_o), the column length (L) and the viscosity of the carrier gas (η) (see Section 5.2):

$$B_o = 2\eta\varepsilon L \frac{p_o}{p_i^2 - p_o^2} u_o \qquad \text{eq.4.22b}$$

Assuming a value of $\varepsilon = 0.40$ (see Section 4.2), eq.4.22a can be approximated as:

$$B_o \approx \frac{d_p^2}{1012} \qquad \text{eq.4.23}$$

The specific permeability of **open-tubular columns** can be established from eq.4.22b, by taking $\varepsilon = 1.00$ and substituting the term expressed by the Hagen-Poiseuille equation (see Section 5.2) for u_o:

$$u_o = \frac{d_c^2}{64\eta L} \cdot \frac{p_i^2 - p_o^2}{p_o} = \frac{r_c^2}{16\eta L} \cdot \frac{p_i^2 - p_o^2}{p_o} \qquad \text{eq.4.24}$$

where d_c and r_c are the inner diameter and radius of the column tube, respectively. The resulting expression for open-tubular columns is:

$$B_o = \frac{d_c^2}{32} = \frac{r_c^2}{8} \qquad \text{eq.4.25}$$

Tables IVa–b list specific permeability values for packed columns containing three different sizes of supports and for open-tubular columns with five different inner diameters, calculated using eqs.4.22a ($\varepsilon = 0.40$), 4.25 and 4.26a–b.

4.11 Flow Resistance Parameter

This term (Φ) is a reduced, dimensionless form, comparing packing density and permeability of columns packed with different particles:

$$\Phi = \frac{d_p^2}{B_o} \qquad\qquad \text{eq.4.26a}$$

In the case of well-packed columns the value of Φ will remain constant. This can be seen in Table IVa, which lists the values of the flow resistance parameter for the packed columns: It is 1013.95 ± 5.5 ($\pm 0.5\%$).

In the case of open-tubular columns we take (as in other reduced terms) the inner diameter of the column (d_c) instead of the particle diameter:

$$\Phi = \frac{d_c^2}{B_o} \qquad\qquad \text{eq.4.26b}$$

On the other hand, we have seen that in the case of open-tubular columns:

$$B_o = \frac{d_c^2}{32} \qquad\qquad \text{eq.4.25}$$

Therefore, in the case of open-tubular columns the flow resistance parameter has a fixed value: It is equal to $\Phi = 32$, as shown in Table IVb, page 33.

Mesh size	Average particle diameter d_p μm	Specific permeability B_o $\times 10^{-7}$ cm^2	Flow resistance parameter Φ
60/80	230	5.22	1013.4
80/100	163	2.62	1014.0
100/120	137	1.85	1014.5

Table IVa. Specific permeability (B$_o$) values of packed columns containing three different particle size supports and their flow resistance parameters (Φ)*

---

* calculated using eqs. 4.22a ($\varepsilon = 0.40$) and 4.26a.

Column tube diameter d_c mm	Specific permeability B_o $\times 10^{-7}$ cm^2	Flow resistance parameter Φ
0.10	31	32
0.25	195	32
0.32	320	32
0.53	878	32
0.75	1758	32

Table IVb. Specific permeability (B$_o$) values of open-tubular columns with five different diameters and their flow resistance parameter (Φ)*

4.12 Diffusion in the Stationary Phase

In various basic relationships related to the chromatographic separation process one needs the diffusion coefficients of the analyte in the mobile (D_M) and stationary (D_S) phase of the column.

Questions related to the diffusion coefficient in the **mobile phase** are discussed in Section 5.6.

Very little information is available concerning diffusion in the **stationary phases**. Kong and Hawkes[3] presented empirical equations to calculate the D_S values of n-alkanes in silicone phases, as a function of the carbon number (c_n) of the n-alkane and the column temperature. For SE-30 methylsilicone the empirical equation describing D_S has the following form:

$$\ln D_S = -8.870 + 0.3836 c_n + \frac{446.1 - 498.1 c_n}{R \cdot T_c} \qquad \text{eq.4.27}$$

where T_c is the absolute column temperature. The same authors also described a generalized empirical equation for the diffusion coefficient of n-alkanes in phenyl-methyl silicones as a function of the carbon number of the n-alkane, the density (ρ) and phenyl content (%Ph) of the phase and the absolute temperature (T_c):

---

* calculated using eqs 4.25, and 4.26b.

	SE-30	OV-7	OV-25
	methyl silicone	phenyl-methyl silicone	
Phenyl content, mole-%	-	20	75
Density, g/mL	0.980	1.021	1.150
	D_S [x10^{-7} cm^2/sec]		
n-Hexane	588	184	65
n-Heptane	463	141	50
n-Octane	365	109	39
n-Nonane	288	83	30
n-Decane	227	64	23
n-Undecane	179	49	18

Table V. Diffusion coefficients of n-alkanes in three silicone phases at 130°C, calculated using the equations of Kong and Hawkes.

$$\ln D_S = -8.162 + 0.421c_n - 0.0135(\% Ph) + \frac{1144 - 548.9c_n - 1815\rho}{R \cdot T_c}$$

eq.4.28

In both eqs.4.27 and 4.28, R, the gas constant, is 1.987 cal/(deg.-mole).

Table V lists D_S values of n-alkanes at 130°C, in SE-30 methylsilicone and two (OV-7 and OV-25) phenyl-methyl silicone phases, calculated using eqs.4.27 and 4.28.

4.13 References

[1] J. Schoenmakers, *JHRC/CC* **11**, 278–282 (1988).

[2] Cs. Horváth, in *The Practice of Gas Chromatography* (L.S. Ettre and A. Zlatkis, eds.), Interscience Publishers, New York, 1967; p.196.

[3] J.M. Kong and S.J. Hawkes, *J. Chromatogr. Sci.* **14**, 279–287 (1976).

PART V

The Mobile Phase (Carrier Gas)

In gas chromatography the mobile phase is an inert gas, the so-called *carrier gas*. In this context "inert" not only means that the carrier gas does not react with the sample components, but also that it has no role in the retardation (sorption-desorption, partition) process. In this, the situation differs from that in liquid chromatography. Helium, hydrogen and nitrogen are used in most cases as the carrier gas.

The carrier gas and its flow through the column is characterized by a number of values.*

5.1 Volumetric Flow Rate

In practice, this value (expressed as mL/min) is usually measured at column outlet, with a soap bubble flow meter. In this way, one obtains F, the flow rate at column outlet and ambient temperature, under wet gas conditions. This value has to be corrected to dry gas conditions (F_a), in order to eliminate

---

* The best concise discussion of the basic relationships associated with the carrier gas flow in GC columns is by Guiochon [1].

Temperature °C	0.0	0.2	0.4	0.6	0.8
16	13.634	13.809	13.987	14.166	14.347
17	14.530	14.715	14.903	15.092	15.284
18	15.477	15.673	15.871	16.071	16.272
19	16.477	16.685	16.894	17.105	17.319
20	17.535	17.735	17.974	18.197	18.422
21	18.650	18.880	19.113	19.349	19.587
22	19.827	20.070	20.316	20.505	20.815
23	21.068	21.324	21.583	21.845	22.110
24	22.377	22.648	22.922	23.198	23.476
25	23.756	24.039	24.826	24.617	24.912

Table VI. Vapor pressure of water, in torr (Hg mm)

the effect of water vapor in the bubble flow meter, and then to the flow rate at column temperature (F_c):

$$F_a = F \frac{p_a - p_w}{p_a} \qquad \text{eq.5.1}$$

$$F_c = F_a \frac{T_c}{T_a} = F \frac{p_a - p_w}{p_a} \cdot \frac{T_c}{T_a} \qquad \text{eq.5.2}$$

where T_a and T_c are the ambient and column temperatures, respectively (both in degrees Kelvin), p_a is the ambient (barometric) pressure and p_w is the partial pressure of water at ambient temperature (both expressed in the same units). Table VI on this page lists the values of p_w at temperatures between 16.0 and 25.8°C.

5.2 Linear Mobile Phase Velocity

In theoretical treatments concerning column efficiencies, the linear mobile phase velocity (expressed as cm/s) is used rather than the flow rate. Three different velocity values can be expressed.

5.2.1 Velocity at Column Outlet

The mobile phase velocity at column outlet (u_o) can be related directly to the flow rate F_c:

$$u_o = F_c / A_c \qquad\qquad \text{eq.5.3}$$

where A_c is the cross-sectional area of the column tubing available to the mobile phase (see Section 4.2).

Although eq.5.3 gives a convenient way to establish u_o, in practice one usually calculates it from the value of the average linear gas velocity ($\overline{u}$) (see Section 5.2.2):

$$u_o = \overline{u} / j \qquad\qquad \text{eq.5.4}$$

where j is the gas compression correction factor (see Section 5.3, below). There are two reasons for using this indirect way to calculate u_o. The first is that in GC, $\overline{u}$ is the more important expression and, as discussed below, it can be easily determined from measuring the holdup time of an unretained analyte. The second reason is that in the case of packed columns, the establishment of the cross-sectional area of the column tubing available to the moving phase is based on approximation.

For an **open-tubular column** the outlet velocity can be expressed by the Hagen-Poiseuille equation:

$$u_o = \frac{d_c^2}{64\eta L}\cdot\frac{p_i^2 - p_o^2}{p_o} = \frac{r_c^2}{16\eta L}\cdot\frac{p_i^2 - p_o^2}{p_o} \qquad\qquad \text{eq.5.5a}$$

where d_c and r_c are the column tube's inner diameter and radius respectively and L is its length; η is the carrier gas viscosity at column temperature, and p_i and p_o are the inlet and outlet pressures, respectively.

Eq.5.5 can also be written in the following form:

$$u_o = \frac{d_c^2 p_o}{64\eta L}(P^2 - 1) = \frac{r_c^2 p_o}{16\eta L}(P^2 - 1) \qquad\qquad \text{eq.5.5b}$$

where P is the relative pressure:

$$P = p_i / p_o \qquad\qquad \text{eq.5.6}$$

In this and other similar calculations one must remember that a system of homogenous dimensions must be used. For example, if viscosity is in poise ($g \cdot cm^{-1} \cdot s^{-1}$), then length and diameter must be in cm and pressure in dyne·cm^{-2} ($g \cdot cm^{-1} \cdot s^{-2}$). Similarly, if viscosity is given in pascal·s ($kg \cdot m^{-1} \cdot s^{-1}$), then length and diameter must be in m and pressure in pascal ($kg \cdot m^{-1} \cdot s^{-2}$). For convenience's sake, Table VII on page 39 presents the conversion factors of pressure units.

It should be emphasized that p_i is the absolute inlet pressure. Gauges on gas chromatographs do not measure the inlet pressure but the pressure drop along the column (Δp):

$$\Delta p = p_i - p_o \qquad \text{eq.5.7}$$

The value of u_o describes the gas velocity at column outlet. In GC, however, this value does not describe the true conditions in the column. This is due to the compressibility of the carrier gas.

5.2.2 Average Linear Gas Velocity

Because of gas compressibility, the density, pressure and velocity will be different at each point in the column. The average linear gas velocity, ($\overline{u}$), can be obtained from the velocity measured at column outlet by applying the gas compression correction factor (j: See Section 5.3):

$$\overline{u} = u_o j \qquad \text{eq.5.8}$$

A convenient way to directly measure the average linear gas velocity is to determine the time that is necessary for an unretained analyte to pass through the column. Except for a few adsorption columns, air can be used as a typical unretained compound. Thus, if the detector responds to air (e.g., it is a thermal-conductivity detector) then one simply injects a small air sample into the instrument and establishes the time elapsed from the instant of sample introduction to the maximum of the air peak. In the case of detectors that do not respond to air (e.g., the flame-ionization detector), one can utilize an organic substance which has almost no retardation on the column. Methane may be assumed to fulfill this criterion.

	atm	bar	dyne/cm^2	g/cm^2	psi	pascal
atm	1	1.01325	1.01325 x10^6	3323x10^3	14.6959	1.01325x10^5
bar	0.98692	1	1.00000 x10^6	1.01972x10^3	14.5038	1.0000 x10^5
dyne/cm^2	0.98692x10^{-6}	1.00000 x10^{-6}	1	1.01971x10^{-3}	1.45038 x10^{-5}	1.0000x10^{-5}
g/cm^2	0.96784 x10^{-3}	0.98067x10^{-3}	0.98067x10^{-3}	1	0.014223	0.98067x10^2
psi	0.068046	6.89476x10^{-2}	6.89476x10^4	70.307	1	6.89476x10^3
pascal	9.86692x10^{-6}	1.00000x10^{-5}	10.0000	1.01971x10^{-5}	1.45038x10^{-4}	1

To convert a value expressed in the unit in the left-hand vertical column to a unit given in the top horizontal row, multiply the value by the factor indicated. For example, a pressure of 5 atm is equal to 5 x 14.6959 = 73.4795 psi and 5 x 1.01325 x 10^5 = 5.06625 x 10^5 pascal.

Table VII. Pressure conversion factors

Since in the time measured the unretained solute passes the length of the column (one usually neglects the extracolumn volumes), the average linear gas velocity ($\overline{u}$) can be calculated by dividing column length (L) by the retention time of the unretained solute (air or methane, t_M):

$$\overline{u} = L / t_M \qquad \text{eq.5.9}$$

In Section 2.2 we have discussed ways to calculate the gas holdup time from the retention time of three members of a homologous series with equidistant carbon numbers. Naturally, the value of t_M calculated in this way can also be used in eq.5.9 to calculate the average linear carrier gas velocity.

Combining eq.5.9 with eq.3.5 we obtain the familiar relationship expressing the total retention time as a function of column length, average linear gas velocity and the retention factor (capacity ratio, k):

$$t_R = \frac{L}{\overline{u}}(k+1) \qquad \text{eq.5.10}$$

This means that in the case of identical columns (same stationary phase, same phase ratio and same temperature), assuming a constant value for $\overline{u}$, the total retention time is a direct function of column length.

5.2.3 Reduced Mobile Phase Velocity

The *reduced mobile phase velocity* (v) is a dimensionless term used mainly when comparing various chromatographic techniques. It compares the flow velocity along the column with the speed of molecular diffusion. The latter is expressed as the ratio of the diffusion coefficient of the analyte in the mobile phase (D_M; see Section 5.6) to the particle size (d_p: in the case of packed columns) or to the inner tube diameter (d_c: in the case of open-tubular columns). Thus, the reduced velocity can be expressed as:

$$v = u \cdot d_p / D_M \qquad \text{(packed columns)} \qquad \text{eq.5.11}$$

$$v = u \cdot d_c / D_M \qquad \text{(open-tubular columns)} \qquad \text{eq.5.12}$$

In eqs.5.11 and 5.12, *u* represents the mobile phase velocity in general. Because of the compressibility of gases, the average linear gas velocity ($\overline{u}$) is to be used in gas chromatography.

The reduced velocity concept was first introduced in 1963, by Giddings [2]. It is widely used in liquid chromatography but is equally valid in gas chromatography. In Sections 11.4.7 and 11.6 we shall discuss some utilization of the reduced velocity.

5.3 Compression (Compressibility) Correction Factor

The carrier gas *compression (compressibility) correction factor* (*j*) was first described by James and Martin [3] in their fundamental paper on gas-liquid partition chromatography. Its value can be calculated from the relative pressure (*P*; see eq.5.6):

$$j = \frac{3}{2} \cdot \frac{P^2 - 1}{P^3 - 1} \qquad \text{eq.5.13}$$

Eq.5.13 is the most widely used form of the compression correction factor. Based on the general algebraic expressions for (a^2-b^2) and (a^3-b^3):

$$a^2 - b^2 = (a - b)(a + b) \qquad \text{eq.5.14}$$

$$a^3 - b^3 = (a - b)(a^2 + ab + b^2) \qquad \text{eq.5.15}$$

it can also be written in the following form:

$$j = \frac{3}{2} \cdot \frac{P + 1}{P^2 + P + 1} \qquad \text{eq.5.16}$$

Table VIII on page 42 lists values of *j* as a function of *P* and $\Delta p / p_o$.

5.4 Pressure Correction Factors

These pressure correction factors are used in GC calculations. These are needed because of the pressure gradient along the column.

P	$\Delta p/p_o$	j	j'	j"
1.0	0	1.0000	1.0000	1.0000
1.5	0.5	0.7895	0.9868	1.0128
2.0	1.0	0.6429	0.9643	1.0332
2.5	1.5	0.5385	0.9423	1.0510
3.0	2.0	0.4615	0.9231	1.0651
3.5	2.5	0.4030	0.9067	1.0759
4.0	3.0	0.3571	0.8928	1.0842
4.5	3.5	0.3204	0.8811	1.0906
5.0	4.0	0.2903	0.8710	1.0957
5.5	4.5	0.2653	0.8622	1.0998
6.0	5.0	0.2492	0.8547	1.1031
∞	∞	0	0.7500	1.1250

Table VIII. Values of j, j' *and* j" *for different values of* P *and* $\Delta p/p_o$

5.4.1 Pressure Correction Factor of Halász

This factor is used when relating the pressure drop (Δp) to the average gas velocity (see below). Its value can be expressed as:

$$j' = \frac{3}{4} \cdot \frac{(P^2-1)\,(P+1)}{P^3-1} \qquad \text{eq.5.17a}$$

where P is the relative pressure ($= p_i/p_o$). Again, eq.5.17a can be modified by expressing (P^2-1) and (P^3-1) based on eqs.5.14 and 5.15, respectively:

$$j' = \frac{3}{4} \cdot \frac{(P+1)^2}{P^2+P+1} = \frac{3}{4} \cdot \frac{P^2+2P+1}{P^2+P+1} \qquad \text{eq.5.17b}$$

The reduced pressure correction factor was first described by Halász, Hartmann and Heine [4], who wrote it in the following form:

$$j' = \frac{3}{4} \times \frac{4+4(\Delta p/p_o)+(\Delta p/p_o)^2}{3+3(\Delta p/p_o)+(\Delta p/p_o)^2} \qquad \text{eq.5.18}$$

In all these equations the pressure drop (Δp) is expressed as:

$$\Delta p = p_i - p_o \qquad \text{eq.5.19}$$

where p_i and p_o are the inlet and outlet pressures of the column, respectively.

Table VIII on page 42 lists values of j' as a function of P and $\Delta p/p_o$. As seen, it ranges from unity at $P = 1$ to 0.7500 at $P = \infty$.

5.4.2 Pressure Correction Factor of Giddings

This factor (j'') is used in the Giddings equation describing the relationship of column efficiency (HETP) from various parameters when considering the outlet velocity (see Section 11.4.6). Its value can be expressed as:

$$j'' = \frac{9}{8} \cdot \frac{(P^4 - 1)(P^2 - 1)}{(P^3 - 1)^2} \qquad \text{eq.5.20}$$

This factor was first described by Stewart, Seager and Giddings [5].

Table VIII on page 42 lists values of j'' as a function of P and p/p_o. As seen, it ranges from unity at $P = 1$ to $j'' = 1.1250$ at $P = \infty$.

5.5 Pressure Drop Along the Column

The pressure drop (Δp) along the column is necessary to maintain a mobile phase flow:

$$\Delta p = p_i - p_o \qquad \text{eq.5.19}$$

As already mentioned earlier, p_i is the absolute inlet pressure. The pressure drop can be related to column parameters and the mobile phase viscosity (η) and velocity (u):

$$\Delta p = \frac{L \cdot \eta \cdot u}{B_o} \qquad \text{eq.5.21}$$

where L is the column length and B_o is the specific permeability of the column (see Section 4.10). In gas chromatography, u refers to the average linear carrier gas velocity, $\bar{u}$.

Eq.5.21 would indicate that the pressure drop increases linearly with velocity. This is true only if the mobile phase is incompressible. If it is compressible — such as in gas chromatography — then the reduced pressure correction factor (j', see Section 5.4) must be applied to Δp:

$$\Delta p \cdot j' = \frac{L \cdot \eta \cdot \overline{u}}{B_o}$$
eq.5.22

This indicates that in such a case the pressure drop increases slightly faster than the average linear gas velocity. At low relative pressure values the difference is small (cf. Table VIII, page 42); at high relative pressure values the maximum difference is 25%.

As we have seen (cf. Section 4.10), for open-tubular columns the specific permeability can be expressed as:

$$B_o = d_c^2 / 32 = r_c^2 / 8$$
eq.4.25

Thus, for such columns eq.5.22 can be written as:

$$\Delta p \cdot j' = \frac{8L \cdot \eta \cdot \overline{u}}{r_c^2}$$
eq.5.23

If we substitute $u_o \cdot j$ for $\overline{u}$ (cf.eq.5.8):

$$\Delta p (j'/j) = \frac{8L \cdot \eta \cdot u_o}{r_c^2}$$
eq.5.24

Dividing eq.5.17 by eq.5.13 we obtain:

$$\frac{j'}{j} = \frac{2 \times 3}{3 \times 4} \frac{(P^3-1)\,(P^2-1)\,(P+1)}{(P^2-1)\,(P^3-1)} = \frac{P+1}{2}$$
eq.5.25

and therefore:

$$\frac{\Delta p (P+1)}{2} = \frac{8L \cdot \eta \cdot u_o}{r_c^2}$$
eq.5.26

5.6 Diffusion in the Mobile Phase

In the calculation of the reduced velocity (see Section 5.4) and also of the reduced film thickness (see Section 4.6) the diffusion coefficient of the analyte in the mobile phase (D_M) is needed; it is also used in the Van Deemter-Golay equations (see Section 11.4). There are two empirical expressions used to calculate this value.

5.6.1 Diffusion in a Gaseous Mobile Phase

In the case of a gaseous mobile phase (gas chromatography) the diffusion coefficient of the analyte can be calculated by using the Fuller-Schettler-Giddings equation [6,7]:

$$D_{AB} = \frac{0.001T^{1.75}(1/M_A + 1/M_B)^{1/2}}{p\left[(\Sigma v_A)^{1/3} + (\Sigma v_B)^{1/3}\right]^2} \qquad \text{eq.5.27}$$

The meaning of the individual symbols is explained below, together with the units used in the calculation:

D_{AB} diffusion coefficient of analyte A in gas B (cm^2/s)

T absolute temperature (K)

M molecular mass of the analyte (A) and the gas (B) (g)

p pressure (atm)

Σv the sum of the so-called atomic volume increments, for the analyte (A) and the gas (B)

The individual atomic increments for simpler organic compounds are:

C = 16.5

H = 1.98

O = 5.48

Thus , e.g., for n-octane (C_8H_{18}), Σv_A is 8×16.5 + 18×1.98 = 167.64.

Increment values for three other heteroatoms are:

N = 5.69

Cl = 19.5

S = 17.0

However, as pointed out in the original publication [6], these values are based only on a few data points.

For simple gases (either analytes or used as the carrier gas) Table IX on page 46 lists the values proposed for Σv.

Gas	Σv
Air	7.07
Ammonia	14.9
Argon	16.1
Carbon dioxide	26.9
Carbon monoxide	8.9
Hydrogen	7.07
Helium	2.88
Krypton	22.8
Neon	5.59
Nitrogen	17.9
Oxygen	16.6
Water (vapor)	12.7

Table IX. Values of Σv for simple gases to be used in the calculation of the diffusion coefficient (D_M) according to the Fuller-Schettler-Giddings equation (eq.5.27).

Table X on page 47 lists diffusion coefficients (D_M) of n-alkanes at 130°C, in three carrier gases, calculated according to the Fuller-Schettler-Giddings equation.

Concerning the value of D_M, one important comment must be made. As seen in eq.5.27, the value of D_M is inversely proportional to pressure: In fact, for a given system (same analyte, same mobile phase and same temperature), the product $D_M \cdot p$ is constant. In gas chromatography, the pressure along the column changes continuously from p_i at the inlet to p_o at the outlet. This means that the value of D_M is also changing along the column. However, in calculations one is usually calculating its value at outlet pressure, assuming it to be equal to 1 atm. This presents an inaccuracy into calculations involving the value of D_M, such as the Van Deemter-Golay equations (see Section 11.4). In most of the cases, the difference is minor; however, if the pressure drop along the column is significant, one must either calculate the value of D_M at the average pressure of the column (see eq.5.33) or use the Van Deemter-Golay equation according to the modification of Giddings (see Section 11.4.6).

Compound	D_M [cm²/s] in		
	Hydrogen	Helium	Nitrogen
n-Hexane	0.536	0.447	0.135
n-Heptane	0.498	0.412	0.124
n-Octane	0.467	0.383	0.116
n-Nonane	0.440	0.360	0.109
n-Decane	0.418	0.340	0.103
n-Undecane	0.398	0.323	0.098

Table X. Diffusion coefficients (D_M) of n-alkanes at 130°C in three carrier gases, calculated using the Fuller-Schettler-Giddings equation (eq.5.27).*

5.6.2 Diffusion in a Liquid Mobile Phase

Although our subject is only gas chromatography, for the sake of completeness we shall also give the relationship suggested for the calculation of D_M for a liquid mobile phase (i.e., in liquid chromatography). This value may be calculated with help of the so-called Wilke-Chang equation [8]:

$$D_{AB} = \frac{7.4 \times 10^{-8} (\Psi_B M_B)^{1/2} T}{\eta_B V_A^{0.6}}$$ 5.28

The meaning of each symbol is explained below, together with the units used in the calculation:

D_{AB} diffusion coefficient of analyte A in liquid B (cm²/s)

M molecular mass of the analyte (A) and the liquid (B) (g)

T absolute temperature (K)

η_B viscosity of liquid B (cP)

Ψ_B solvent constant (association factor)

V_A molar volume of the analyte molecule (cm³):

   $V_A = M_A / \rho_A$

ρ_A density of analyte A (g/cm³)

---

* Calculated for a pressure of 1 atm.

Suggested values for Ψ_B are:

water: 2.6

methanol: 1.9

ethanol: 1.5

for other non-associating molecules: 1.0

In the case of solvent mixtures, both M_B and Ψ_B are to be calculated from the mole fractions. For detailed rules of such calculations see the literature ([8] or e.g. [9]).

The Wilke-Chang equation is generally valid only for simple molecules with a molar mass of $M_A < 1000$. In the case of macromolecules the actual diffusion coefficients are higher by a factor of 1.3–3.2 than those calculated using eq.5.28.

5.7 Gas Viscosity

Viscosity is a temperature-dependent parameter. When the temperature of a gas is increased, its viscosity also increases. In most gas chromatographs used with open-tubular columns, the inlet pressure (pressure drop) is kept constant. Thus, according to eq.5.21, the product $\eta \cdot \overline{u}$ will be constant. Therefore, in temperature programming, when the viscosity of the gas increases, the average linear gas velocity will correspondingly decrease.

It is difficult to express in absolute terms the viscosity (η) of a gas. However, it is generally accepted that the ratio of its viscosity at two temperatures is proportional to the corresponding temperature ratio:

$$\eta_i / \eta_o = (T_i / T_o)^x \qquad \text{eq.5.29}$$

where the temperatures are in degrees Kelvin. There is no general agreement, however, concerning the value of exponent x: A detailed examination of literature data [10] gave the following values:

helium: $x = 0.646$

hydrogen: $x = 0.680$

nitrogen: $x = 0.725$

Figure 6 on page 49 presents η vs. T plots for the three gases, calculated using these exponent values and considering the viscosity at 0°C (273.15 K) as η_o.

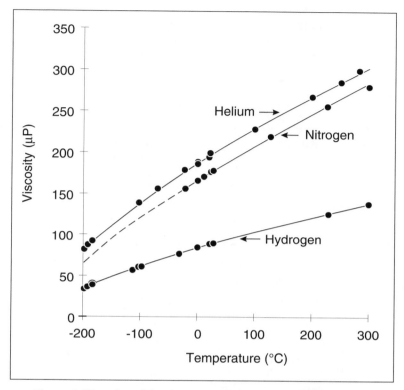

Figure 6. Viscosity of the three most frequently used carrier gases, as a function of temperature. Solid circles correspond to measured values listed in the literature. Plots correspond to values calculated according to eq.5.29 using the exponents given in the text [10].

In a more limited temperature range, one can describe the η vs. T relationship by an empirical linear equation (where T_i is in °C) [10]:

$$\eta_i = aT_i + b \qquad\qquad \text{eq.5.30}$$

Considering the range between 0 and 300°C, the values of a and b are listed in Table XI on page 50. Viscosity calculated according to eq.5.30 is obtained in micropoise (μP). If values in SI units (Pa·s) are desired, multiply the values by 0.1. Table XII on page 50 lists viscosity values of the three gases in the range between 0 and 300°C, calculated according to eq.5.30.

	Helium	Hydrogen	Nitrogen
a	0.3992945	0.3837787	0.1827282
b	186.61685	167.35536	83.98985
r *	0.99979	0.99971	0.99915

Table XI. Constants of eq.5.30 used to calculate the viscosity of three gases, in the temperature range between 0 and 300°C.

Temperature	Viscosity (μP)		
(°C)	Helium	Hydrogen	Nitrogen
0	186.62	83.99	167.36
20	194.60	87.64	175.03
40	202.59	91.30	182.71
60	210.57	94.95	190.38
80	218.56	98.61	198.06
100	226.55	102.26	205.73
120	234.53	105.02	213.41
140	242.52	109.57	221.08
160	250.50	113.23	228.76
180	258.49	116.88	236.44
200	266.48	120.53	244.11
220	274.46	124.19	251.79
240	282.45	127.84	259.46
260	290.43	131.50	267.14
280	298.42	135.15	274.81
300	306.41	138.81	282.49

Table XII. Viscosity values for three gases, in the temperature range of 0-300°C, calculated with help of eq.5.30 [10].

---

*Linear regression according to eq.5.30 calculated for six measured values in the given temperature range.

5.8 Pressure and Velocity Gradients

Owing to the compressibility of the gases, as the pressure in the column decreases, the gas expands: thus, for a constant rate of mass flow, the linear velocity of the carrier gas increases along the length of the column from inlet to outlet. These changes can be calculated as a function of the z/L ratio, where L is the column length and z is the distance from the inlet (see Figure 7, below).

5.8.1 Pressure Gradient

The relative pressure at point z is:

$$P_z = p_z/p_o \qquad\qquad \text{eq.5.31}$$

where $p_i > p_z > p_o$. P_z can be derived as a function of $P = p_i/p_o$ and the relative distance z/L along the column[11]:

$$P_z = \sqrt{P^2 - [(z/L)(P^2 - 1)]} \qquad\qquad \text{eq.5.32}$$

Figure 8 on page 52 shows plots of P_z vs. z/L for various P values; the corresponding numerical data are listed in Table XIII on page 53. As seen, a significant drop in column pressure occurs only in the very last segment of the column, particularly at relatively low P values. For example, if we assume atmospheric pressure ($p_o = 1$ atm) and $P = 2$ ($p_i = 2$ atm), then at $z/L = 0.9$ (i.e., at a distance equalling $0.9L$), the relative pressure is still $P_z = 1.14$.

The *average pressure in the column* ($\overline{p}$) can be calculated

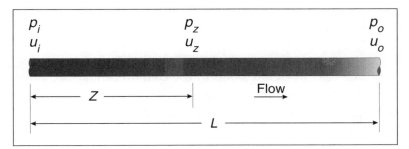

Figure 7. Conditions in the column

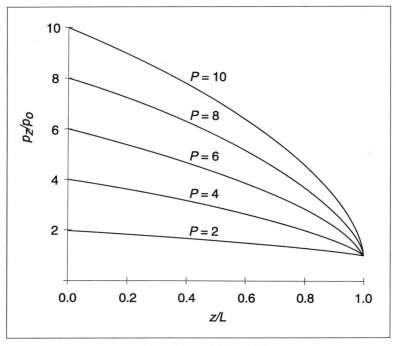

Figure 8. Plots of $P_z = p_z/p_o$ *vs.* z/L *for* $P = p_i/p_o$ *values between* $P = 2$ *and* $P = 10$

from the outlet pressure (p_o) and the compression correction factor (j):

$$\overline{p} = p_o/j \qquad\qquad\text{eq.5.33}$$

5.8.2 Velocity Gradient

Because of gas compression the linear velocity (u) of the carrier gas changes in the column: It is lowest at the inlet (u_i) and the highest at the outlet (u_o):

$$u_i < u_z < u_o$$

The velocity at any point in the column can be calculated from the corresponding P_z value (see Section 5.8.1): one calculates j_z, the compression correction factor at that point:

z/L	Relative pressure for the whole column (p_i/p_o)				
	2.00	4.00	6.00	8.00	10.00
	Value of p_z/p_o				
0.00	2.00	4.00	6.00	8.00	10.00
0.20	1.84	3.61	5.39	7.17	8.96
0.40	1.67	3.16	4.69	6.23	7.77
0.60	1.48	2.65	3.87	5.12	6.37
0.80	1.26	2.00	2.83	3.69	4.56
0.90	1.14	1.58	2.12	2.70	3.30
0.95	1.07	1.32	1.66	2.04	2.44
1.00	1.00	1.00	1.00	1.00	1.00

Table XIII. Pressure along the column, as a function of the p_i/p_o value

$$j_z = \frac{3}{2}\frac{P_z^2 - 1}{P_z^3 - 1}$$
eq.5.34

which is equal to the ratio of the velocity at that point (u_z) to the velocity at column outlet (u_o):

$$u_z/u_o = j_z$$
eq.5.35

Figure 9 on page 54 plots u_z/u_o vs. z/L for various P values; the numerical values are listed in Table XIV on page 54. The evaluation of these plots is similar to the evaluation of the P_z vs. z/L plots: The velocity changes relatively little in most of the column and the most significant increase occurs in a relatively small segment toward the end of the column.

We have already seen the expression for the average linear carrier gas velocity in the column:

$$\overline{u} = u_o j$$
eq.5.8

where j is calculated for $P = p_i/p_o$.

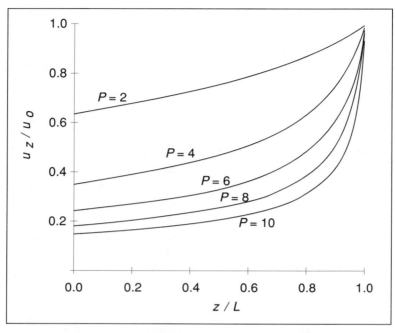

Figure 9. Plots of u_z/u_o *vs.* z/L *for* $P = p_i/p_o$ *values between* $P = 2$ *and* $P = 10$

z/L	Relative pressure for the whole column (p_i/p_o)				
	2.00	4.00	6.00	8.00	10.00
	Value of u_z/u_o				
0.00	0.6429	0.3571	0.2442	0.1849	0.1486
0.20	0.6842	0.3920	0.2704	0.2057	0.1656
0.40	0.7337	0.4411	0.3083	0.2355	0.1903
0.60	0.7965	0.5130	0.3681	0.2839	0.2306
0.80	0.8811	0.6429	0.4853	0.3843	0.3165
0.90	0.9332	0.7624	0.6146	0.5050	0.4246
0.95	0.9658	0.8566	0.7368	0.6332	0.5493
1.00	1.0000	1.0000	1.0000	1.0000	1.0000

Table XIV. Velocity along the column, as a function of the p_i/p_o *value*

5.9 References

[1] G. Guiochon, *Chromatogr. Rev.* **8**, 1–47 (1966).

[2] J.C. Giddings, *Anal. Chem.* **35**, 1338–1341 (1963).

[3] A.T. James and A.J.P. Martin, *Biochem. J.* **50**, 679–690 (1952).

[4] I. Halász , K. Hartmann and E. Heine, in *Gas Chromatography 1964 (Brighton Symposium)* (A. Goldup, ed.), Institute of Petroleum, London, 1965; pp.38–61.

[5] G.H. Stewart, S.L. Seager and J.C. Giddings, *Anal. Chem.* **31**, 1738 only (1959).

[6] E.N. Fuller and J.C. Giddings, *J. Gas Chromatogr.* **3**, 222–227 (1965).

[7] E.N. Fuller, P.F. Schettler and J.C. Giddings, *Ind. Eng. Chem.* **58**, 19–27 (May 1966).

[8] C.R. Wilke and P. Chang, *Am. Inst. Chem. Eng. J.* **1**, 264–270 (1955).

[9] V.R. Meyer, *Praxis der Hochleistungs-Flüssigchromatographie.* Salle+Sauerländer Verlag, Frankfurt am Main/Aarau, 6th edition 1990; pp.101–103.

[10] L.S. Ettre, *Chromatographia* **18**, 243–248 (1984).

[11] L.S. Ettre, *JHRC/CC* **10**, 637–640 (1987).

PART VI

Retention Volumes

Retention characteristics can be expressed not only in time but also in volume units. Such terms are used relatively rarely in the general GC practice. However, these terms are widely applied in the theory of GC, as well as in detailed treatments of the technique, and in liquid chromatography.

Retention volumes are established by multiplying the respective retention time values with the volumetric flow rate measured at column outlet and corrected to column temperature (F_c). Because of gas compression, additional terms also exist in gas chromatography which do not have equivalent time units that can be directly measured from the chromatogram. For convenience, however, such time expressions are derived indirectly from the corresponding retention volume terms. •

6.1 Mobile Phase Holdup Volume

We have seen the meaning of the mobile phase holdup time (t_M) in Section 2.1: it is the retention time of a solute that is not retained by the stationary phase and thus is traveling through the column with the speed of the carrier gas. The retention volume corresponding to the mobile phase holdup time is the *mobile phase holdup volume* or, in GC, the *gas holdup volume* (V_M):

$$V_M = t_M \cdot F_c \qquad \text{eq.6.1}$$

6.2 Corrected Gas Holdup Volume

As mentioned above, F_c is determined at column end pressure (ambient pressure). In order to establish the real conditions in a GC column, the compression correction factor (j; see Section 5.3) must be applied:

$$V_M^o = t_M \cdot F_c \cdot j = V_M \cdot j \qquad \text{eq.6.2}$$

V_M^o is called the *corrected gas holdup volume*. Its physical meaning can be explained by the following consideration. We have seen earlier (cf. eq.5.3) that, in general, the flow rate is related to the cross-sectional area of the column available to the carrier gas (A_c) and the outlet gas velocity (u_o) by:

$$F_c = u_o A_c \qquad \text{eq.6.3}$$

In an open-tubular column, A_c is equal to the cross-sectional area of the column tube, while in a packed column the interparticle porosity (ε) must be considered (cf. eqs.4.3a and 4.5):

$$A_c = r_c^2 \pi \qquad \text{(open-tubular columns)}$$

$$A_c = \varepsilon \cdot r_c^2 \pi \qquad \text{(packed columns)}$$

We have also seen (cf. eq.5.8 and 5.9) that:

$$\overline{u} = u_o \cdot j$$
$$\overline{u} = L / t_M$$

where $\overline{u}$ is the average linear gas velocity and L is the column length. Substituting these expressions into eq.6.3 we can write that:

$$F_c \cdot t_M \cdot j = A_c \cdot L \qquad \text{eq.6.4}$$

The right-hand side of eq.6.4 is equal to the gas-phase volume of the column (V_G), while the left-hand side of this equation has already been described in eq.6.2; thus:

$$V_M^o = V_M \cdot j = V_G \qquad \text{eq.6.5}$$

In other words, V_M^o is equal to the gas-phase volume of the column.

In this derivation extracolumn dead volumes (V_e) were neglected: It was assumed that all the volumes occupied by the carrier gas between sample introduction and the detection point are in the column. Actually,

$$V_M^o = V_G + V_e \qquad \text{eq.6.6}$$

However, in a well-designed GC system, $V_e \ll V_G$ and thus V_e may be neglected.*

In some books the symbol V_M is used for what is here called V_M^o; correspondingly, the symbol V_A is used to express the gas holdup volume. In the present treatment the official I.U.P.A.C. nomenclature is followed, according to which the sign ^{o} should be used to express correction for gas compression.

6.3 Total Retention Volume

The *total retention volume* of an analyte (V_R) is the carrier gas volume corresponding to the total retention time (t_R):

$$V_R = t_R \cdot F_c \qquad \text{eq.6.7}$$

6.4 Corrected Retention Volume

The *corrected retention volume* (V_M^o) is the carrier gas volume obtained from the total retention volume when correcting for gas compression:

$$V_R^o = V_R \cdot j \qquad \text{eq.6.8}$$

---

* When using an open-tubular column, comparison of V_G (calculated from the dimensions of the column) and of V_M^o (calculated from the measured gas holdup time) can indicate the extent of any extracolumn volume.

Substituting eq.6.7 for V_R we can write that:

$$V_R^o = t_R \cdot j \cdot F_c \qquad \text{eq.6.9}$$

Ad analogiam to the other terms, the *corrected retention time* (t_R^o) may be derived from this expression by dividing it by the flow rate:

$$t_R^o = \frac{V_R^o}{F_c} = \frac{t_R \cdot j \cdot F_c}{F_c} = t_R \cdot j \qquad \text{eq.6.10}$$

One can, however, debate whether this term, representing a time corrected for gas compression, has any real meaning.

It should be noted that in liquid chromatography where the compression of the mobile phase is negligible, $j = 1$. Therefore, there the total and corrected retention volumes (and times) are identical.

6.5 Adjusted Retention Volume

The *adjusted retention volume* of the analyte (V_R') is the carrier gas volume corresponding to the adjusted retention time (t_R'):

$$V_R' = t_R' \cdot F_c \qquad \text{eq.6.11}$$

Expressing t_R' from eq.2.1 we can write eq.6.11 in the following way:

$$V_R' = (t_R - t_M) \, F_c = t_R F_c - t_M F_c \qquad \text{eq.6.12}$$

Substituting the proper expressions from eqs.6.1 and 6.7 we obtain:

$$V_R' = V_R - V_M \qquad \text{eq.6.13}$$

6.6 Net Retention Volume

The *net retention volume* of an analyte (V_N) is the carrier gas volume obtained from the adjusted retention volume (V_R'), correcting it for gas compression:

$$V_N = V_R' \cdot j = t_R' \cdot j \cdot F_c \qquad \text{eq.6.14}$$

Again, we can modify this equation by expressing t_R from eq.2.1:

$$V_N = (t_R - t_M)\,j \cdot F_c = t_R \cdot j \cdot F_c - t_M \cdot j \cdot F_c \qquad \text{eq.6.15}$$

Substituting the proper terms from eqs.6.2 and 6.9 we obtain:

$$V_N = V_R^o - V_M^o \qquad \text{eq.6.16}$$

Logically, the net retention volume should have the symbol $V_R^{o\prime}$ because superscript ' indicates correction for gas holdup, and superscript ^{o} indicates correction for gas compression. The symbol V_N has been adapted in order to avoid the need for a double superscript.

Again, the *net retention time* (t_N) may be derived from the net retention volume by dividing it by the flow rate:

$$t_N = \frac{V_N}{F_c} = \frac{(t_R - t_M)\,j \cdot F_c}{F_c} = (t_R - t_M)\,j = t_R' \cdot j \qquad \text{eq.6.17}$$

One can, however, again debate whether such a term has a real meaning.

It should be noted that in liquid chromatography where the compression of the mobile phase is negligible, $j = 1$, and the adjusted and net retention volumes (and times) are identical.

6.7 Specific Retention Volumes

The specific retention volumes are derived from the net retention volume, and they express it for unit amount (1 gram) of the stationary phase: the subscript g in the symbol of the term refers to this. Two versions of this term exist.

The *specific retention volume of an analyte at column temperature* (V_g^θ) has no temperature correction since the flow rate used in the calculation of the net retention volume (V_N) was already corrected to column temperature:

$$V_g^\theta = V_N / W_S \qquad \text{eq.6.18}$$

where W_S is the amount (mass) of the stationary phase present in the column. The dimension of V_g^θ is always mL/g.

The *specific retention volume of an analyte at 0°C* (V_g) corrects the value of V_g^θ to 0°C:

$$V_g = V_g^\theta \frac{273.15}{T_c} = \frac{V_N}{W_S} \cdot \frac{273.15}{T_c}$$ eq.6.19

where T_c is the absolute column temperature (in K).

The logarithm of the specific retention volume V_g can be expressed as a function of the heat of solution (ΔH_S) of the analyte in the stationary phase:

$$\log V_g = -\frac{\Delta H_S}{2.3 R \cdot T_c} + a$$ eq.6.20

where R is the gas constant, a is a constant and T_c is the absolute column temperature. Eq.6.20 is usually solved graphically, by plotting V_g against the reciprocal absolute temperature ($1/T_c$).

PART VII

Column Efficiency

We have seen in Part I how the separation in a column takes place. The degree of separation of the individual sample components will naturally depend on the characteristics of the column. In practical gas chromatography it is important to be able to express the "goodness" of the column and to be able to relate it to actual separation problems. In this part we shall investigate the various terms expressing column efficiency, i.e., the "goodness" of the column. In Part VIII we will deal with the relationships expressing the chromatographic separation.

It should be noted that the plate number and plate height terms are always calculated for isothermal operation.

7.1 Number of Theoretical Plates (Plate Number)

As mentioned in Part I, the width of the chromatographic peak is a function of band-spreading in the column. This band-spreading is a function of many complex processes that will be discussed in Part XI, dealing with the fundamental relationships of chromatography. Too-wide peaks can reduce the resolution of closely-spaced peaks; therefore, the goal is to have as little band-spreading as possible. Since band-spreading in a given column is also a function of time, we measure the "goodness" of a chromatographic column by an expression relating the retention time to peak width. The term was borrowed from the science of distillation and is called the *number of theoretical plates* (*N*).* In the fundamental expression of the plate number, peak width is characterized by its standard deviation (σ; see Section 2.3):

$$N = (t_R / \sigma)^2 \qquad \text{eq.7.1}$$

where t_R is the total retention time of the analyte.

In practice we prefer to use values that can be directly measured in the chromatogram. These are the peak width at base (w_b), the peak width at half height (w_h) or the peak width at inflection points (w_i), all of which are a direct function of the standard deviation. Substituting for σ the proper expressions from eqs.2.11a-c, we obtain the well-known equations for the number of theoretical plates (plate number):

$$N = 16(t_R / w_b)^2 \qquad \text{eq.7.2a}$$

$$N = 5.545(t_R / w_h)^2 \qquad \text{eq.7.2b}$$

$$N = 4(t_R / w_i)^2 \qquad \text{eq.7.2c}$$

In eq.7.2b, 5.545 is equal to $\left(2\sqrt{2\ln 2}\right)^2 = 8\ln 2$.

The plate number is a dimensionless number; the larger *N*, the more efficient the column.

The number of theoretical plates depends on the column length, among other variables. We have seen earlier (see Section 5.2.2) that under identical conditions, the retention time is a linear function of column length. Since in the plate number equations we have the second power of the total retention time, the number of theoretical plates is proportional to the square of column length; this means that one needs four times a given length to double the plate number.

---

* In older nomenclatures the symbol *n* was recommended for the plate number and *N* for the effective plate number (see Section 7.3); however, their use was not consistent. Therefore, the new I.U.P.A.C. nomenclature uses *N* for the plate number, indicating the effective plate number by a subscript (N_{eff}).

7.2 Height Equivalent to One Theoretical Plate (Plate Height, HETP)

Since the number of theoretical plates depends on column length, terms were devised that permit comparison of column efficiencies independent of their lengths. Such a term is the *height equivalent to one theoretical plate* or *plate height* (*HETP*) (*H*), once again an expression borrowed from distillation theory:

$$H = L/N \qquad\qquad \text{eq.7.3}$$

where *L* is the column length. The HETP is usually expressed in cm or mm: the smaller the HETP, the more efficient the column.

The physical meaning of the HETP can be explained by going back to Figure 2 on page 6: It is the length of the small segments in which equilibrium conditions were assumed to exist during the passage of the sample through the column. It is easy to understand that the more such equilibrium steps are in a given column length, the better is the separation of the sample components from each other.

Both the number of theoretical plates and the HETP were established in 1941 by Martin and Synge for column liquid chromatography [1], in order to explain the phenomena occurring during separation in the column. The same concept was also utilized in 1952 by James and Martin when first describing gas-liquid partition chromatography [2]. Actually, equilibrium never happens in the column, and thus no "theoretical plates" exist. Still, the plate height (and plate number) concept was retained because it represents a useful way to explain the complex process occurring in the chromatographic column.

The concept of the number of theoretical plates and the HETP can also be derived from a different consideration. A Gaussian distribution peak is characterized by its variance (σ^2). The total variance of a peak emerging from a column can be considered as the sum of the variances contributed by the length elements of the column. Assuming that the column is uniform, the variance of the analyte's distribution at the end of the column is directly proportional to the column length (*L*):

$$\sigma^2 = H \cdot L \qquad \text{eq.7.4}$$

We call the proportionality factor the plate height or HETP.

Eq.7.4 can be modified in the following way:

$$H = \frac{\sigma^2}{L} = \left(\frac{\sigma}{L}\right)^2 \cdot L \qquad \text{eq.7.5}$$

Here both σ and L are given in **length** units. Going to **time** units for the term in parentheses, we can replace L with t_R, the total retention time, and σ with the standard deviation of the peak, now expressed in time units. Thus, eq.7.5 can be written as:

$$H = \left(\frac{\sigma}{t_R}\right)^2 \cdot L = \frac{L}{(t_R / \sigma)^2} \qquad \text{eq.7.6}$$

and

$$(t_R / \sigma)^2 = L / H \qquad \text{eq.7.7}$$

L/H tells us how many theoretical plates (H) are present in a column having a length of L. We call it the number of theoretical plates or plate number (N):

$$N = (t_R / \sigma)^2 = L / H \qquad \text{eq.7.8}$$

which is the same relationship as already given in eqs.7.1 and 7.3.

7.3 Effective Plate Number and Effective Plate Height

In eq.7.1, describing the number of theoretical plates, the total retention time (t_R) was used. We have seen earlier (cf. eq.2.1) that the total retention time is composed of two parts: the gas holdup time (t_M) and the adjusted retention time (t'_R). Thus we can write eq.7.1 in the following form:

$$N = \left(\frac{t_M + t'_R}{\sigma}\right)^2 \qquad \text{eq.7.9}$$

Of the two terms in the numerator, the gas holdup time

does not contribute to the column's efficiency because it only represents the time necessary for the carrier gas to pass through the column. For example, if one were to connect a long uncoated capillary tube before the actual column, t_M would be larger and thus a higher apparent theoretical plate number value would be obtained, although the actual performance of the column remained unchanged.

In order to eliminate this discrepancy, a modified plate number term has been introduced: It is called the *number of effective plates* (N_{eff}):

$$N_{eff} = (t'_R / \sigma)^2 \qquad\qquad \text{eq.7.10}$$

Substituting the proper peak widths we obtain:

$$N_{eff} = 16 (t'_R / w_b)^2 \qquad\qquad \text{eq.7.11a}$$

$$N_{eff} = 5.545 (t'_R / w_h)^2 \qquad\qquad \text{eq.7.11b}$$

$$N_{eff} = 4 (t'_R / w_i)^2 \qquad\qquad \text{eq.7.11c}$$

From the number of effective plates we can also derive an expression similar to the HETP: It is called the *height equivalent to one effective plate* (H_{eff}):

$$H_{eff} = L/N_{eff} \qquad\qquad \text{eq.7.12}$$

In books and publications the expressions "number of theoretical effective plates" and "height equivalent to one effective theoretical plate" (HEETP) have also been used to describe these two terms. This is incorrect, however, since plate number and plate height are either theoretical or effective, but cannot be both.

The effective plate number (and height) concept was first suggested in 1959-1960 by Purnell [3,4]; the terms were first used in 1961, by Desty et al. [5].

The relationship between N and N_{eff} can be established by expressing t_M from eq.3.1 and substituting it into eq.7.9. The resulting relationship is described by the following equation:

$$N = N_{eff}\left(\frac{k+1}{k}\right)^2 \qquad \text{eq.7.13a}$$

And conversely:

$$N_{eff} = N\left(\frac{k}{k+1}\right)^2 \qquad \text{eq.7.13b}$$

Similarly, we can derive the relationship between H and H_{eff}:

$$H = H_{eff}\left(\frac{k}{k+1}\right)^2 \qquad \text{eq.7.14a}$$

$$H_{eff} = H\left(\frac{k+1}{k}\right)^2 \qquad \text{eq.7.14b}$$

7.4 Reduced Plate Height

This term was originally introduced by Giddings [6-7] and popularized by Knox for liquid chromatography [8]. The reduced plate height (h) relates the HETP (H) to the particle diameter (d_p) or, in the case of open-tubular columns, to the inner tube diameter (d_c):

$$h = H / d_p \qquad \text{(packed columns)} \qquad \text{eq.7.15a}$$

$$h = H / d_c \qquad \text{(open-tubular columns)} \qquad \text{eq.7.15b}$$

As seen, the reduced plate height expresses the HETP as a multiple of the particle diameter or the tube diameter:

$$H = h \cdot d_p \qquad \text{(packed columns)} \qquad \text{eq.7.16a}$$

$$H = h \cdot d_c \qquad \text{(open-tubular columns)} \qquad \text{eq.7.16b}$$

In Sections 11.4.7 and 11.6 we shall discuss some utilizations of the reduced plate height.

7.5 References

[1] A.J.P. Martin and R.L.M. Synge, *Biochem. J.* **35**, 1358–1368 (1941).

[2] A.T. James and A.J.P. Martin, *Biochem. J.* **50**, 679–690 (1952).

[3] J.H. Purnell, *Nature (London)* **164**, 2009 only (1959).

[4] J.H. Purnell, *J. Chem. Soc.* **1960**, 1268–1274.

[5] D.H. Desty, A. Goldup and W.T. Swanton, in *Gas Chromatography (1961 Lansing Symposium)* (N. Brenner, J.E. Callen and M.D. Weiss, eds.), Academic Press, New York, 1962; pp.105–135.

[6] J.C. Giddings, *Anal. Chem.* **35**, 1338–1341 (1963).

[7] J.C. Giddings, *Dynamics of Chromatography, Part I.* M. Dekker, Inc., New-York, 1965; p.58.

[8] J.N. Done, J.H. Knox and J. Loheac, *Applications of High-Speed Liquid Chromatography*, J. Wiley & Sons, London, 1974.

PART VIII

Separation and Resolution

Until now we investigated only a single peak in the chromatogram. However, chromatography is a separation technique; thus, our interest is in how the separation of consecutive peaks can be expressed and related to column performance.

8.1 Relative Retention

The position of each peak in the chromatogram is influenced by many factors. In GC their relative position is primarily an indication of their selective interaction with the stationary phase. The more selective a phase is toward one analyte vs. another, the further apart their peaks will be in the chromatogram. This selectivity can be best expressed by the ratio of their distribution constants (partition coefficients), or other values directly proportional to the partition coefficient.

Depending on our point of view, there are two versions of this term.

8.1.1 Separation Factor

This term is used if the relative retardation of two *adjacent* peaks is expressed. It is indicated by the symbol α:

$$\alpha = \frac{K_2}{K_1} = \frac{k_2}{k_1} = \frac{t'_{R2}}{t'_{R1}} = \frac{V'_{R2}}{V'_{R1}} \qquad \text{eq.8.1}$$

where K is the partition coefficient, k is the retention factor (capacity ratio) and t_R' and V_R' are the adjusted retention time and volume, respectively. By definition, $t_{R2}' > t_{R1}'$, i.e., the value of α is always larger than (or equal to) unity.

In Part I we gave the thermodynamic description of the partition coefficient (cf. eq.1.3). If we describe this for the two analytes ($t_{R2}' > t_{R1}'$; $K_2 > K_1$) and substitute the corresponding expressions into eq.8.1, we can write that:

$$\alpha = \frac{K_2}{K_1} \approx \frac{p_1^o}{p_2^o} \cdot \frac{\gamma_1^o}{\gamma_2^o} \qquad \text{eq.8.2a}$$

or, in logarithmic form:

$$\log \alpha \approx \log(p_1^o/p_2^o) + \log(\gamma_1^o/\gamma_2^o) \qquad \text{eq.8.2b}$$

where p^o refers to the saturated vapor pressure of the analyte and γ^o is the activity coefficient of the analyte at infinite dilution in the stationary phase. An "approximately equal to" sign is used in eq.8.2 because, strictly speaking, this relationship is exact only for infinitely dilute solutions of the analytes in the stationary phase.

Eq.8.2b clearly indicates the two contributions to the separation factor: the first term containing the saturated vapor pressures is related to the analytes and depends only on temperature, while the second term containing the activity coefficients depends on the selective properties of the stationary phase. This thermodynamic meaning of the separation factor was first described by Herington [1].

The separation factor is widely used in the fundamental relationships expressing chromatographic separation and resolution, usually as a variation of $\alpha / (\alpha\text{-}1)$ (see Section 11.3). Table XV on page 71 lists some numerical values that are plotted in Figure 10, page 72.

Earlier (see Section 2.2) we indicated that for a homologous series analyzed under identical conditions there is a

α	$\left(\dfrac{\alpha}{\alpha-1}\right)^2$	$\left(\dfrac{\alpha-1}{\alpha}\right)$
1.01	10,201.00	0.00990
1.02	2,601.00	0.01961
1.03	1,178.78	0.02913
1.04	676.00	0.03846
1.05	441.00	0.04762
1.06	312.11	0.05660
1.07	233.65	0.06542
1.08	182.25	0.07410
1.09	146.68	0.08257
1.10	121.00	0.09091
1.15	58.78	0.13043
1.20	36.00	0.16667
1.25	25.00	0.20000
1.30	18.78	0.23077

Table XV. Numerical values of some important functions of α

semi-logarithmic relationship between the carbon number (c_n) of members of the homologous series and their adjusted retention time (t_R'):

$$\log t_R' = a \cdot c_n + b \qquad\qquad \text{eq.2.3}$$

If we write eq.2.3 for two consecutive homologs with carbon numbers z and $z+1$:

$$\log t_{R(z+1)}' = a \cdot (z+1) + b$$

$$\log t_{Rz}' = a \cdot z + b$$

and then subtract the second equation from the first, we obtain that:

$$\log t_{R(z+1)}' - \log t_{Rz}' = a$$

or

$$\log(t_{R(z+1)}' / t_{Rz}') = a$$

where a is a constant (the slope of the semi-logarithmic plot corresponding to eq.2.3). If the logarithm of a value is constant,

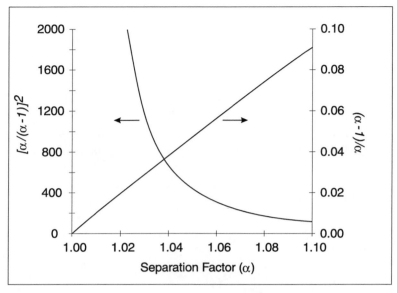

Figure 10. Plots of $[\alpha / (\alpha\text{-}1)]^2$ and $[(\alpha\text{-}1) / \alpha]$ against α

then the value itself is also a constant:

$$\alpha = t'_{R(z+1)} / t'_{Rz} = const.$$ eq.8.3

In other words, the separation factor of two consecutive members of a homologous series is constant. This is an important conclusion often used in GC calculations.

8.1.2 Relative Retention

We have seen that the separation factor is a fundamental value depending only on the phase, the two analytes and the column temperature. Because of this, it can also be used for another purpose: If we fix the second peak and express the retention of various analytes relative to this standard, then the obtained values will be characteristic of the particular analytes. In this usage we call the term the *relative retention* and identify it with the symbol r:

$$r = \frac{K_i}{K_{st}} = \frac{k_i}{k_{st}} = \frac{t'_{Ri}}{t'_{R(st)}} = \frac{V'_{Ri}}{V'_{R(st)}}$$ eq.8.4

Subscript *i* refers to the particular analyte and subscript *st* refers to the standard peak.

Because we have now fixed one peak and express the retention of all the other peaks relative to that peak, the value of *r* may be smaller, larger or equal to unity, depending on the position of the analyte peaks relative to the peak of the standard.

The relative retention term is one of the bases of qualitative analysis.

8.1.3 Temperature Dependency of Relative Retention

We have seen that both the distribution constant (partition coefficient: K) and the retention factor (capacity ratio: k) can be described by Antoine-type equations (cf. eqs.1.4 and 3.3a). A similar relationship exists also between the separation factor (α) or the relative retention (r) and the absolute column temperature (T_c):

$$\log \alpha = \frac{a}{T_c} + b \qquad\qquad \text{eq.8.5}$$

Similarly to K and k, the $\log \alpha$ vs. T_c relationship can also be approximated in a relatively short temperature range by a linear equation:

$$\log \alpha = a' \cdot T_c + b' \qquad\qquad \text{eq.8.6}$$

Depending on the value of a' (the slope of the plot), three possibilities exist. In most of the cases, the value of a' is negative; this means that decreasing the temperature will increase the value of relative retention. The second possibility is that $a' \approx 0$. In this case, the relative retention remains practically constant regardless of column temperature. Finally, the third possibility is that $a' > 0$. In such cases the relative retention will decrease when column temperature is lowered.

8.2 Unadjusted Relative Retention

This term is obtained by relating the total retention times (volumes) of two analytes:

$$\alpha_G = \frac{t_{R2}}{t_{R1}} = \frac{V_{R2}}{V_{R1}} = \frac{k_2 + 1}{k_1 + 1} \qquad \text{eq.8.7a}$$

$$r_G = \frac{t_{R2}}{t_{R1}} = \frac{V_{R2}}{V_{R1}} = \frac{k_2 + 1}{k_1 + 1} \qquad \text{eq.8.7b}$$

Subscript G commemorates E. Glueckauf, who first used this expression [2]. The symbols *RRT* (for *relative retention time*) or α^* have also been used to describe the unadjusted relative retention values.

As mentioned, the values of the relative retention or the separation factor are — for a given stationary phase and at a given temperature — constant and independent of the type and dimension of the column and of the carrier gas flow rate. However, this is not true about the unadjusted relative retention. Therefore, such values cannot be applied to another instrument or column (even if prepared with the same stationary phase) and cannot be directly compared with tables in which true relative retention (separation factor) values are listed. On the other hand, in one laboratory, in a given chromatographic system (instrument, column) and under specified analytical conditions, such values still can be reproduced very well.

The values of r and r_G or α and α_G are interrelated [3]. Expressing k_1 from eq.8.1 and substituting it into eq.8.7a, we obtain:

$$\alpha_G = \frac{k_2 + 1}{(k_2 / \alpha) + 1} = \frac{\alpha(k_2 + 1)}{k_2 + \alpha} \qquad \text{eq.8.8a}$$

Similarly we can derive that:

$$\alpha_G = \frac{\alpha \cdot k_1 + 1}{k_1} \qquad \text{eq.8.8b}$$

From these equations we can also express α:

$$\alpha = \frac{\alpha_G \cdot k_2}{(k_2 + 1) - \alpha_G} \qquad \text{eq.8.9a}$$

$$\alpha = \frac{\alpha_G(k_1 + 1) - 1}{k_1} \qquad \text{eq.8.9b}$$

Thus, when the retention factors (capacity ratios) are known, the value of α (or r) can be calculated from α_G (r_G) and *vice versa*.

8.3 Peak Resolution

In a given stationary phase, the relative retention of two consecutive peaks is constant at a given column temperature and is independent of the column type and dimensions. However, the actual degree of separation of these two peaks will depend on the efficiency of the column: the higher the column efficiency, the better the resolution of the two peaks. This is illustrated in Figure 11 on page 76: The two peaks have the same relative retention but their resolution is quite different.

The **resolution** of two consecutive peaks is expressed by relating the distance between the two peak maxima (Δt) to the average width of the peaks. A number of expressions exist depending on whether peak width is characterized by the standard deviation (σ), the peak width at half height (w_h) or the peak width at base (w_b). The official I.U.P.A.C. nomenclature uses the latter; thus, *peak resolution* (R_s) is expressed as (see Figure 12, page 76):

$$R_s = \frac{t_{R2} - t_{R1}}{\dfrac{w_{b1} + w_{b2}}{2}} = \frac{2 \cdot \Delta t}{w_{b1} + w_{b2}} \qquad \text{eq.8.10}$$

where $\Delta t = t_{R2} - t_{R1} = t'_{R2} - t'_{R1}$. In practice, when considering two closely spaced peaks, we may assume that $w_{b1} = w_{b2}$. Therefore:

$$R_s \approx \Delta t / w_{b2} \qquad \text{eq.8.11.}$$

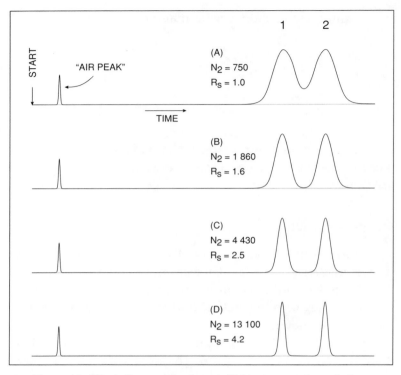

Figure 11. The influence of column efficiency on peak resolution. Constant values: $t_M = 90$ *sec,* $t_{R1} = 792$ *sec,* $t_{R2} = 930$ *sec,* $t_{R1}' = 702$ *sec,* $t_{R2}' = 840$ *sec;* $\alpha = 1.2$.

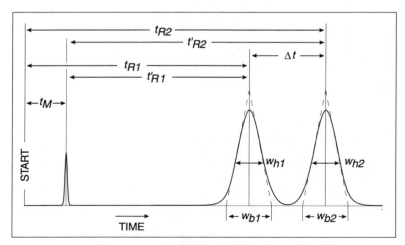

Figure 12. Values measured from a chromatogram containing two closely spaced peaks

Using the relationship expressed in eqs.8.10 and 8.11, a value of $R_s = 1.0$ means 4σ-resolution and about 94% separation of the two peaks, while at $R_s = 1.5$ (6σ-resolution: "baseline resolution") the separation is practically complete. Figure 13, below, illustrates these two cases, together with the case when $R_s = 1.177$ (4.7σ-resolution), which corresponds to the definition of the separation number (see Section 8.4.2).

The essential problem with the peak resolution expression is that it requires the value of the peak width at base (w_b) or the value of the standard deviation (σ) of the peak ($w_b = 4\sigma$). On

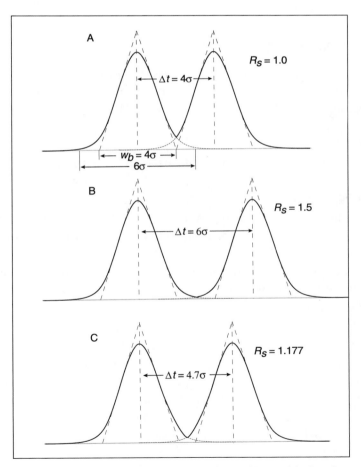

Figure 13. Peak resolution. A = 4σ-resolution ($R_s = 1.0$); B = 6σ-resolution ($R_s = 1.5$); C = 4.7σ-resolution ($R_s = 1.177$).

the other hand, one is rarely measuring w_b directly, since it involves drawing the tangents to the inflection points of the peak, etc. The peak width value that can be directly measured is the peak width at half height (w_h).

Modern data systems that establish the peak area by integration may also calculate σ from the peak area (cf. eq.2.13), assuming a Gaussian peak. Since all peak widths are functions of σ, one can also calculate the peak width at base from the measured value of the peak width at half height (see Section 2.3):

$$w_b = \frac{4}{2\sqrt{2\ln 2}} w_h = 1.699 w_h \qquad\qquad \text{eq.2.12}$$

8.4 Peak Capacity

In the separation of a complex mixture one often asks the question, "How many peaks could be separated in a given time segment in the chromatogram?" The answer to this question depends on the resolution we require between two consecutive peaks.

The situation is illustrated in a simplified form in Figure 14 on page 79. Here, the peaks are indicated by triangles, the bases of which are equal to the peak width at base (w_b), while the distance between the peak maxima is equal to 4σ; as we have seen above, a 4σ-resolution between two consecutive peaks gives a value of $R_s = 1.0$. In this figure the widths at base of the peaks are identical while in the practice they increase with time. However, in the calculations below we shall consider this situation.

The question stated above can be formulated in a different way, by assuming two major peaks, the retention times of which are t_{R1} and t_{R2} ($t_{R2} > t_{R1}$). What we are interested in is the number of peaks we can place between these two major peaks and resolve all of them by a specified resolution. If the distance between two consecutive peak maxima is Δt, then we can place n peaks between the two major peaks:

$$n = \frac{t_{R2} - t_{R1}}{\Delta t} - 1$$

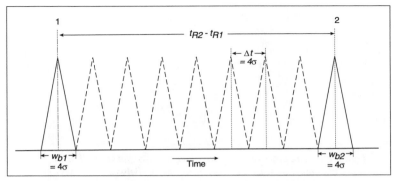

Figure 14. Peaks separated between two major peaks.
$(t_{R2} - t_{R1}) / \Delta t = 8$; $w_{b1} = w_{b2} = 4\sigma$ *and* $\Delta t = 4\sigma$.

This is illustrated in Figure 14, where the fraction $(t_{R2} - t_{R1}) / \Delta t$ is equal to eight, and we can place seven peaks between the two major peaks.

In this example the distance between the maxima of two consecutive peaks (Δt) was equal to their base width (4σ). However, it can also be a different value: we can specify any desired resolution for two consecutive peaks. In this respect, we must distinguish between two resolution values: Resolution R_s is calculated for the two major peaks, while R_s^* is the specified resolution between two consecutive peaks. It can be deduced that the number of peaks that can be resolved by resolution R_s^* between the two major peaks (the *peak number, PN*) can be calculated from the following equation:

$$PN = \frac{R_s}{R_s^*} - 1 \qquad\qquad \text{eq.8.12}$$

Figure 14 gives an example for 4σ-resolution, i.e., when $R_s^* = 1.0$. In this case there is about 94% separation between two consecutive peaks (see Figure 13, page 77). Figure 15 on page 80 illustrates the situation when $R_s^* = 1.5$, which refers to 6σ-resolution, i.e., complete separation between two consecutive peaks. This figure illustrates three cases differing in the actual resolution between the two major peaks. For example, if $R_s = 3$, then:

$$PN = \frac{3}{1.5} - 1 = 1$$

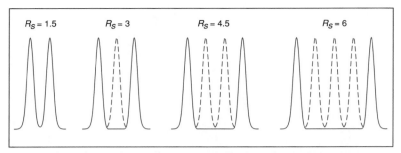

Figure 15. Peak pairs with different degrees of separation. The broken line indicates the peaks that can be separated between the two major peaks (solid line), with a resolution of $R_s^* = 1.5$ between two consecutive peaks.

and we could place one additional peak between the two major peaks. On the other hand, if $R_s = 6$, then:

$$PN = \frac{6}{1.5} - 1 = 3$$

and we could place three additional peaks between the two major peaks.

As mentioned above, Figure 14 simplified the situation by having a constant peak width for each peak, while in a chromatogram the peak widths increase with time. We address this by calculating R_s using eq.8.10 and *not* eq.8.11; in this way we compensate for the increase in peak width by time.

There are two special terms that utilize the peak capacity concept. They essentially differ only in the value of R_s^*. For more details, see [4].

8.4.1 Effective Peak Number

This concept was described by Hurrell and Perry [5] and refers to the case when $R_s^* = 1$. Thus, the effective peak number * (*EPN*) can be calculated as:

$$EPN = R_s - 1 \qquad\qquad \text{eq.8.13}$$

---

* One must be careful not to mix up the effective *peak* number (*EPN*) with the effective *plate* number (N_{eff}: see Section 7.3).

8.4.2 Separation Number (Trennzahl)

This concept was first introduced by Kaiser [6]. The *separation number* (or, in its original German name, *Trennzahl*) is a peak number expression similar to eq.8.10, except that resolution is calculated in a somewhat different way, by using the peak widths at half height (w_h) instead of the peak widths at base. We are using the acronym *SN* or *TZ* to indicate this term:

$$SN = \frac{t_{R(z+1)} - t_{Rz}}{w_{hz} + w_{h(z+1)}} - 1 \qquad \text{eq.8.14}$$

where *z* and *z+1* refer to two consecutive members of the n-paraffin homologous series. Eq.8.14 gives the number of peaks which can be resolved between the two main peaks, having a 4.7σ-resolution between two consecutive peaks ($R_s{}^* = 1.177$; cf. Figure 13C, page 77).

It should be noted that the value of *SN* depends on the peak pair chosen, even within one homologous series. In other words, *SN* will be different between n-heptane and n-hexane as compared to n-pentadecane and n-tetradecane. Therefore, the peak pair selected for testing must always be specified.

The separation number has two advantages in practical use. The first is that from its value, one can easily establish the smallest difference in the retention index values of two consecutive peaks that can still be separated (see Section 9.3). The second advantage is that — besides the separation power (see Section 8.5) — the separation number is the only term describing column performance that can also be applied to programmed-temperature analysis.

8.5 Separation Power

The calculation of peak resolution or the separation number assumes that one can accurately measure the peak width at half height or at base. However, in the case of closely spaced peaks where we have only partial resolution ($R_s < 1$), this is not always possible. Also, while the meaning of peak resolution values above $R_s = 1$ (or of separation number values) is clear, this is not so with resolution values below $R_s = 1$. For such cases, Kaiser [7] introduced a practical, empirical equation he called the *separation power* (in German: *Trennvermögen*). In its calculation (see Figure 16, below), the maxima of the two peaks are connected and then a perpendicular to the base line is drawn at the valley between the two peaks. "Separation power" (δ) is calculated by expressing the length of this perpendicular above the valley point to its intersect with the line connecting the two peak maxima, relative to the total length of this perpendicular:

$$\delta = \frac{F}{G} 100$$ eq.8.15

The value of δ gives the separation of the two peaks as a percentage. If the two peaks are completely separated, then $F = G$, and thus $\delta = 100\%$. Values above 100% are not possible; as mentioned earlier, δ only has meaning in the case of incompletely separated peaks.

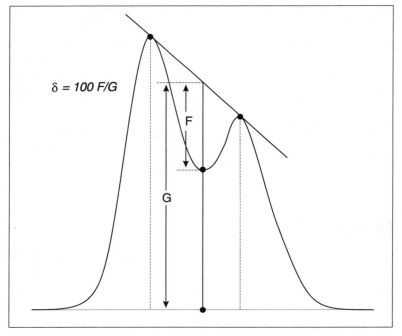

Figure 16. Calculation of the value of "separation power" (δ) of two incompletely resolved peaks

It should be noted that δ also depends on the relative heights of the two peaks, and its value will be higher if the second peak is much smaller than the first. In other words, the "separation power" is an expression directly related to the particular peak pair in the chromatogram. It may be calculated in both isothermal and programmed-temperature operations.

8.6 References

[1] E.F.G. Herington, in *Vapour Phase Chromatography (1956 London Symposium)* (D.H. Desty, ed.), Butterworths, London, 1957; pp. 5–14.

[2] E. Glueckauf, *Trans. Faraday Soc.* **51**, 34–44 (1955).

[3] L.S. Ettre, *J. Chromatogr.* **198**, 229–234 (1980).

[4] L.S. Ettre, *Chromatographia* **8**, 291–299, 355–357 (1975).

[5] R.A. Hurrell and S.G. Perry, *Nature (London)* **196**, 571–572 (1962).

[6] R. Kaiser, *Z. Anal. Chem.* **189**, 1–14 (1962).

[7] R. Kaiser, *Gas-Chromatographie*. Akademische Verlagsgesellschaft, Leipzig, 1960; p.33.

PART IX

The Retention Index System

We have seen that the relative retention expressed as:

$$r = t'_{Ri} / t'_{R(st)}$$

(where subscripts i and st refer to the analyte and a standard, respectively) is a characteristic value for a given stationary phase and temperature and thus is widely used for peak identification.

The relative retention concept, however, has one basic shortcoming: It is almost impossible to fix a single standard. Therefore, published data refer to widely different standards. Also, within one chromatogram, one has the problem of selecting a standard relatively close to the compound of interest, particularly in the case of a multicomponent mixture. If the retention times of the compound of interest and the standard are grossly different, accuracy suffers.

In order to eliminate these problems, Kováts proposed the introduction of the so-called retention index system in 1958.*

---

* The original publications appeared in a German-language journal [1-3]. For detailed discussions on the concept, see refs. [4-8].

9.1 Principles of the Retention Index System

The basic difference between relative retention data and the retention index system is that, in the latter, the retention behavior of a particular analyte is expressed in a uniform **scale** determined by a **series** of closely related standard substances. In this respect it could be compared to our common temperature scale, in which arbitrary numbers are assigned to the temperatures of two specific transitions and the other temperatures are obtained by inter- or extrapolation, using an arbitrary scale (e.g., 100 divisions between the two fixed points). The improvement of the retention index system as compared to our common temperature scale is that extrapolation is never necessary: Every analyte can always be bracketed by two standards.

The fixed points in the Kováts retention index scale are the normal alkanes. The physical meaning of the retention index of an analyte is the following: It is the carbon number (multiplied by 100) of a hypothetical n-alkane that would have the same adjusted retention time as the substance of interest. This is illustrated in Figure 17 on page 86; below we give the mathematical deduction of this expression.

We have already seen that for a homologous series:

$$\log t'_R = a \cdot c_n + b \qquad\qquad \text{eq.2.3}$$

where c_n is the carbon number of the individual homologs (number of carbon atoms in their molecule) and a and b are constants. Let us denote the substance of interest with the subscript i and the carbon numbers of the two bracketing n-alkanes as z and $z+1$:

$$t'_{R(z+1)} > t'_{Ri} > t'_{Rz} \qquad\qquad \text{eq.9.1}$$

Thus, for the three substances we can write:

$$\log t'_{R(z+1)} = a(z+1) + b \qquad\qquad \text{eq.9.2a}$$

$$\log t'_{Ri} = a \cdot x + b \qquad\qquad \text{eq.9.2b}$$

$$\log t'_z = a \cdot z + b \qquad\qquad \text{eq.9.2c}$$

where x is the carbon number of a hypothetical n-alkane that

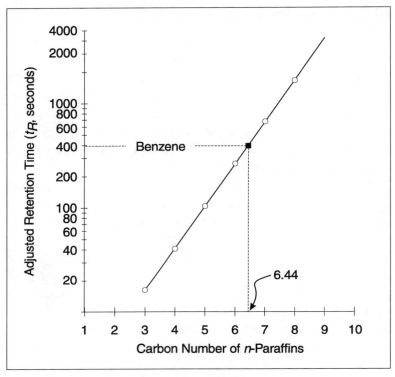

Figure 17. Graphical presentation of the retention index concept. Analyte: benzene. Stationary (liquid) phase: squalane. Column temperature: 60°C. For numerical data see Table XVI on page 87. The carbon number of the hypothetical n-alkane, which would have the same adjusted retention time as benzene, is 6.44; thus, the retention index of benzene on squalane, at 60°C, is 644.

would have the same adjusted retention time as the substance of interest. Subtracting eq.9.2c from eq.9.2b and eq.9.2c from eq.9.2a:

$$\log t'_{Ri} - \log t'_{Rz} = a(x-z) \qquad \text{eq.9.3a}$$

$$\log t'_{(z+1)} - \log t'_{Rz} = a \qquad \text{eq.9.3b}$$

Dividing now eq.9.3a by eq.9.3b we obtain after reorganization that:

$$x = z + \frac{\log t'_{Ri} - \log t'_{Rz}}{\log t'_{R(z+1)} - \log t'_{Rz}} \qquad \text{eq.9.4}$$

Compound	Carbon number	Adjusted retention time seconds
n-Butane	4	41.2
n-Pentane	5	103.9
n-Hexane	6	262.1
n-Heptane	7	668.9
n-Octane	8	1670.0
Benzene	(6.44)	395.4

Table XVI. Numerical data corresponding to Figure 17 on page 86

As mentioned, x is the carbon number of the hypothetical n-alkane that would have the same adjusted retention time as the substance of interest:

$(z+1) > x > z$

It is evident that the value of x would have decimal fractions. Thus, for convenience, we multiply it by 100 to delete the decimal fractions. This value is the Kováts retention index, I, of substance i:

$$I = 100\left[z + \frac{\log t'_{Ri} - \log t'_{Rz}}{\log t'_{R(z+1)} - \log t'_{Rz}}\right] \qquad \text{eq.9.5}$$

It follows from this derivation that the retention index of n-alkanes is always 100 times their carbon number, regardless of the stationary phase and the column temperature; e.g., it is 600 for n-hexane, 700 for n-heptane, etc.

The fraction in the right-hand side of eq.9.5 can be written in the following form:

$$\frac{\log(t'_{Ri}/t'_{Rz})}{\log(t'_{R(z+1)}/t'_{Rz})}$$

The fractions in both the numerator and denominator represent relative retention data. Therefore, the Kováts retention index equation can also be written in the following way:

$$I = 100\left[z + \frac{\log r_{i/z}}{\log r_{(z+1)/z}}\right] \qquad \text{eq.9.6}$$

There are two reasons why the n-alkane series was selected as the basis of the retention index system. First, these substances are available everywhere in a convenient and fairly pure form. The second reason is that n-alkanes are non-polar. Since the retention index of a substance corresponding to the general formula of $C_zH_{(2z+1)}X$ expresses its retention relative to the n-alkane $C_zH_{(2z+2)}$, it is an indication of the influence of group X on retention.

One remark is in order here. As seen earlier, the whole retention index system is based on the validity of eq.2.3, i.e., of the linearity of the $\log t_R'$ vs. c_n plot. In general this plot can be considered linear from about C_4-C_5 on. Therefore, retention index values below 400-500 usually are not utilized in practice.

9.2 Influence of Temperature on the Retention Index

Detailed investigations have shown that the actual relationship between the Kováts retention index (I) and absolute column temperature (T_c) can also be described by an Antoine-type equation:

$$I = A + \frac{B}{T_c + C} \qquad \text{eq.9.7}$$

where A, B and C are constants. In a shorter temperature range the I vs. T_c relationship may be considered linear and therefore written as:

$$I \approx a \cdot T_c + b \qquad \text{eq.9.8}$$

where a and b are again constants. In most cases, $a > 0$ and thus the retention index increases with temperature. Because of this linear relationship, the temperature dependency of the retention index can be described by giving the increment for a given temperature range, usually 10°C.

9.3 Relationship Between the Retention Index and the Separation Number

We have seen earlier (see Section 8.4.2) that the separation number *(SN)* expresses the number of peaks that can be placed between the peaks of two normal paraffins and separated from each other by a resolution of *R = 1.177 (4.7σ-* resolution; cf. Figure 13, page 77):

$$SN = \frac{t_{R(z+1)} - t_{Rz}}{w_{hz} + w_{h(z+1)}} - 1 \qquad \text{eq.8.14}$$

We have also seen that the retention index of n-alkanes is equal to 100 times their carbon number. Thus, the difference between the retention index values of two consecutive n-alkanes is always equal to 100.

Let us now take an example where *SN = 24.* This means that we can have 24 peaks between the two consecutive n-alkanes; thus, the retention index difference between any two adjacent analyte peaks must be *100/(24+1) = 4.* In other words, on this column and within the range defined by the two n-alkanes selected for the determination of the separation number, we can separate by *4.7σ-* resolution two substances, the retention index values of which differ by four index units. We can write this rule in a general way:

$$i = I_y - I_x = \frac{100}{SN+1} \qquad \text{eq.9.9}$$

or

$$SN = \frac{100}{i} - 1 \qquad \text{eq.9.10}$$

where

$$t'_{R(z+1)} > t'_{Ry} > t'_{Rx} > t'_{Rz} \qquad \text{eq.9.11}$$

Here subscripts *y* and *x* indicate the two analytes that we want to separate, and *i* represents the difference in the retention index of these two consecutive peaks that can still be resolved in the specified separation number range with *4.7σ-* resolution.

9.4 Linear Retention Index for Programmed-Temperature Operation

The retention index system is calculated from isothermal data and — as mentioned above — is based on the relationship between the logarithm of the adjusted retention times and the carbon numbers of members of a homologous series.

It has been shown by van den Dool and Kratz [9] that in a temperature-programmed run a similar value can be calculated utilizing direct retention numbers instead of their logarithm. Since both the numerator and denominator contain the difference of two values, here we can use the total retention times. The *linear retention index* (I^T) is described by the following equation:

$$I^T = 100 \left[\frac{t^T_{Ri} - t^T_{Rz}}{t^T_{R(z+1)} - t^T_{Rz}} \right] + z \qquad \text{eq.9.12}$$

where $t_R{}^T$ refers to the retention time measured under programmed-temperature conditions. The *retention temperature*, i.e., the column temperature at which the respective peaks elute from the column, can also be used in the calculation of the linear retention index:

$$I^T = 100 \left[\frac{T_{Ri} - T_{Rz}}{T_{R(z+1)} - T_{Rz}} + z \right] \qquad \text{eq.9.13}$$

where T_R refers to the retention temperature. According to van den Dool and Kratz, the value of I^T is in a reasonable agreement with the isothermal retention index of the same analyte measured at the elution temperature. On the other hand, as stated by Guiochon [10] the agreement is better if the isothermal retention index is measured at the *significant (or equivalent) temperature* introduced by Giddings [11], which is approximately equal to $0.85\ T_R$.

9.5 Other Retention Index Systems

Other similar systems have also been described in the literature. They differ from the Kováts index system in the selection of the standard compounds. From these, the systems intended for the characterization of the methyl esters of branched (and unsaturated) fatty acids are the most important. Here, the methyl esters of straight-chain, saturated fatty acids were proposed to serve as the reference homologous series, expressing the retention of the methyl esters of other fatty acids relative to this series, in a way analogous to the Kováts retention index. Such a system was described almost simultaneously by two groups. Woodford and Gent [12] called the values obtained in this way the *carbon number* of the fatty acids, while Miwa et al. [13] used the term *equivalent chain length*.

9.6 Use of the Retention Index Concept for Stationary Phase Characterization

We have mentioned earlier that the retention index is an indication of the influence of an active group in the analyte molecule on its retention, as compared to the retention of the nearest-eluting normal alkane. A similar approach would be to compare the retention of a given polar compound — containing a certain active group — on a specified liquid phase to its retention on a non-polar phase. It is known that on a non-polar phase the analytes emerge in the order of their boiling point, while their retention on a polar phase depends on the interaction between the active groups in the analytes' molecules and those of the phase. Thus, if the same analyte is analyzed at the same temperature on both a non-polar and a polar phase, then the difference in its retention index values measured on the two phases:

$$\Delta I = I^{polar} - I^{nonpolar} \qquad \text{eq.9.14}$$

is characteristic to this interaction: the larger ΔI the more "polar" is the phase with respect to that particular analyte or to the active group in its molecule. This means that by selecting certain test substances the "polarity" of the phases could be established.

Naturally, it is not enough to select only one test substance because each phase will have different "polarity" toward the various active groups. The question is how many such test substances are needed to adequately characterize a stationary (liquid) phase. This question was investigated in detail by Rohrschneider [14,15], who established that the use of five test substances is adequate for this purpose. Each test substance represents a certain solute group; thus, knowing the values of the retention increments calculated according to eq.9.14, one can predict the applicability of a phase for the separation of a certain compound class.

Rohrschneider's original aim was twofold: to characterize stationary phases and to be able to predict the retention index of an analyte. Thus, his original system consisted of two sets of characteristic values: factors *(solute constants)* characterizing the analyte and factors *(phase constants)* characterizing the stationary phase. However, the first application is very complicated and has never been used in general practice.

Rohrschneider's system was improved by McReynolds [16], who was concerned only with the second application: characterization of stationary phases with respect to their ability to resolve compounds belonging to certain groups.

The Rohrschneider-McReynolds system has its shortcomings, and other systems based on thermodynamic values have been proposed by a number of scientists. However, none of these can match the simplicity of the Rohrschneider-McReynolds system, and thus have not been accepted by the general public. The Rohrschneider-McReynolds system is still used almost universally for stationary phase characterization, and the corresponding phase constants are included in the stationary phase listings of practically every chromatography supply house.

9.6.1 Rohrschneider Constants

Rohrschneider was the first person to establish a system for the characterization of liquid phases on the basis of retention index increments, calculated similarly to eq.9.14. In his system these increments were divided by 100; the resulting values are called the *Rohrschneider constants*. The first test substance is benzene, and the symbol for the first Rohrschneider constant is x. Squalane was selected as a typical non-polar phase; thus, the value of the Rohrschneider constant x is calculated as:

$$x = \frac{\Delta I_{benzene}}{100} = \frac{I_{benzene}^{phase} - I_{benzene}^{squalane}}{100} \qquad \text{eq.9.15}$$

Calculation of the other Rohrschneider constants is done in a similar way. Table XVII, below, lists the test substances proposed by Rohrschneider, the respective symbols of the constants and the substance groups characterized by them.

The practice of dividing retention index increments by 100 comes from Rohrschneider's original aim to (also) predict the retention index of an analyte on a given phase and to the correlation of the phase-specific constants with the solute-specific constants. The reason for the alphabetically irregular order of the symbols is that originally Rohrschneider considered the use of only three *(x, y, z)* phase constants. Subsequently, he realized the need for two more constants but did not want to change the symbols of constants for which values had already been published.

Rohrschneider constants are calculated from retention index measurements at 100°C. Retention index values are temperature-dependent; therefore, the Rohrschneider constants also will depend somewhat on temperature. However, this will influence only the second decimal place, and thus for all practical purposes, the effect of temperature on the Rohrschneider constants is negligible.

Symbol	Test substance	Substance group characterized by the constant
x	benzene	aromatics, olefins
y	ethanol	alcohols, nitriles, acids; mono-, di-, and trichloroalkanes
z	methyl ethyl ketone	ketones, ethers, aldehydes, esters, epoxides and dimethylamino derivatives
u	nitromethane	nitro and nitrile derivatives
s	pyridine	pyridine, dioxane

Table XVII. The Rohrschneider constant system for liquid phase characterization in gas chromatography

9.6.2 McReynolds Constants

The McReynolds constants [16] are based on the same principles as the Rohrschneider constants. The differences are:

- three of the original Rohrschneider test substances were changed to higher homologs;
- since they are used exclusively for liquid phase characterization, division by 100 was omitted;
- retention index measurements were carried out at 120°C;
- a number of additional test substances were also proposed.

The reason for using higher homologs is that the respective original Rohrschneider test substances had such low retention index values on most columns that the bracketing n-alkanes would have to be gases, which of course cannot be mixed directly with a liquid sample. In addition — as mentioned before — this very low end of the $log\ t_R'$ vs. c_n plot is not entirely linear.

McReynolds' original idea for using more than five test substances was to better characterize the phase. However, it was found that for most applications, the first five test substances (and constants) are satisfactory for stationary phase characterization.

Table XVIII on page 95 lists the test substances proposed by McReynolds. The first five test substances are equivalent to the original Rohrschneider test substances, and these are the ones that are used today. In order to distinguish them from the Rohrschneider phase constants, the use of prime (x', etc.) after the symbol was proposed.

As an illustration, Table XIX on page 97 lists the McReynolds constants of methyl, phenyl-methyl and cyanopropyl-phenyl silicones [17]. The table also illustrates that these values are remarkably constant for a given phase composition and are independent of the manufacturer. This can be seen in the case of both the phenyl-methyl and cyanopropyl phenyl silicones, in which data for some phases with the same chemical composition but manufactured by two suppliers are given.

The usefulness of the McReynolds constants in characterizing stationary phases is illustrated in Figure 18 on page 96,

Symbol	Test substance	Substance group characterized by the constant
x'	benzene	aromatics, olefins
y'	butanol-1	alcohols, nitriles, acids
z'	methyl n-propyl ketone	ketones, ethers, aldehydes, esters, epoxides and dimethylamino derivatives
u'	nitropropane	nitro and nitrile derivatives
s'	pyridine	pyridine, dioxane
	2-methylpentanol-2	branched-chain compounds, particularly alcohols
	1-iodobutane	halogenated compounds
	2-octyne	
	1,4-dioxane	
	cis hydrindane	

Table XVIII. The McReynolds constant system for liquid phase characterization in gas chromatography

which plots the McReynolds phase constant x' (which is characteristic for aromatics) against the phenyl content of the phenylmethyl silicones listed in Table XIX. As seen, the plot is remarkably linear.

9.6.3 General Polarity of a Phase

The Rohrschneider-McReynolds phase constants present a way to express the polarity of a phase with respect to a certain compound group. In addition, the sum of the five McReynolds constants is sometimes used as an expression of the general polarity of a phase:

$$P = x'+y'+z'+u'+s' \qquad \text{eq.9.16}$$

Another way to define the "polarity" of a phase may be to use the mean of the first five McReynolds constants:

$$\overline{P} = \frac{x'+y'+z'+u'+s'}{5} \qquad \text{eq.9.17}$$

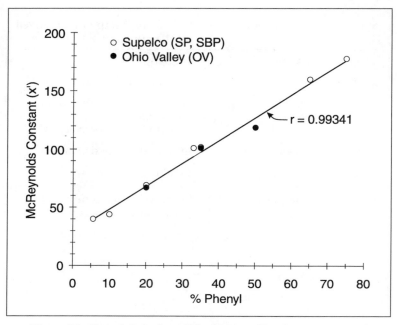

*Figure 18. Plot of the value of the McReynolds phase constant x'
against the phenyl content (in mole-%) of various phenyl-methyl
silicone phases. For numerical data see Table XIX on page 97.*

McReynolds, in his paper, listed the stationary phases according to increasing $\overline{P}$ values, but did not give the actual numerical values of $\overline{P}$.

It should be noted that the meaning of such polarity expressions is less clear than that of the individual phase constants, and thus their usefulness is limited.

9.6.4 Retention of n-Alkanes

As already mentioned, the retention index of n-alkanes is set by definition: It is 100 times their carbon number, regardless of the stationary phase. Therefore, the Rohrschneider-McReynolds phase constants cannot provide any information on the retention of n-alkanes. In order to partly overcome this problem McReynolds, in his original listing [17], presented two additional pieces of information for each phase: the relative re-

Trade name	Composition of the silicone	x'	y'	z'	u'	s'
OV-1	Dimethyl (gum)	16	55	44	65	42
OV-101	Dimethyl (fluid)	17	57	45	67	43
OV-73	Phenyl (5.5%) methyl	40	86	76	114	85
OV-3	Phenyl (10%) methyl	44	86	81	124	88
OV-7	Phenyl (20%) methyl	69	113	111	171	128
SPB-20	Phenyl (20%) methyl	67	116	117	174	131
OV-61	Phenyl (33%) methyl	101	143	142	213	174
OV-11	Phenyl (35%) methyl	102	142	145	219	178
SPB-35	Phenyl (35%) methyl	101	146	151	219	202
OV-17	Phenyl (50%) methyl	119	158	162	243	202
SP-2250	Phenyl (50%) methyl	119	158	162	243	202
OV-22	Phenyl (65%) methyl	160	188	191	283	253
OV-25	Phenyl (75%) methyl	178	204	208	305	280
SP-2300	Cyanopropyl (50%) phenyl	316	495	446	637	530
SILAR 5CP	Cyanopropyl (50%) phenyl	319	495	446	637	531
SP-2310	Cyanopropyl (75%) phenyl	440	637	605	840	670
SILAR 7CP	Cyanopropyl (75%) phenyl	440	638	605	844	673
SP-2330	Cyanopropyl (95%) phenyl	490	725	630	913	778
SILAR 9CP	Cyanopropyl (95%) phenyl	489	725	631	910	778
SP 2340	Cyanopropyl (100%)	520	757	659	942	800
SILAR 10CP	Cyanopropyl (100%)	523	757	659	942	801

Table XIX. McReynolds constants of silicone phases * [17]

* The percentage values represent mole-% values. For example, both compounds given below (Ph=phenyl; Me=methyl):

```
Ph  Me              Ph  Ph
 |   |               |   |
-Si-O-Si-O-         -Si-O-Si-O-
 |   |               |   |
Ph  Me              Me  Me
```

would be listed as containing 50% phenyl groups, disregarding the composition of the monomeric units within the polymer molecule. Actually, the polymer on the right consists of phenyl-methyl-siloxane units, while the polymer on the left consists of diphenylsiloxane and dimethylsiloxane units in 1:1 ratio.

The manufacturers of the individual phases are: **OV** : Ohio Valley Specialty Chemical Co., Marietta, OH; **SP** and **SPB** : Supelco, Inc., Bellefonte, PA; **SILAR** : Applied Science/Alltech, Deerfield, IL.

tention of two consecutive n-alkanes and the slope of the:

$$\log t'_R = a \cdot c_n + b \qquad \text{eq.2.3}$$

relationship for n-alkanes, both at 120°C (t_R' is the adjusted retention time of the n-alkane and c_n is the number of carbon atoms in its molecule).* The slope of this equation is an indication of the solvation power of the phase for a methylene group, while the relative retention of two consecutive n-alkanes is constant on a given phase and at a given temperature. We have already derived this in Section 8.1; accordingly, for two consecutive n-alkanes with z and $z+1$ carbon atoms in their molecule we can write that:

$$\log \frac{t'_{R(z+1)}}{t'_{Rz}} = a \qquad \text{eq.9.18}$$

where a is the slope of eq.2.3. It follows from eq.9.18 that the relative retention (r) of two consecutive n-alkanes is constant on a given column and at a given temperature:

$$r = \frac{t'_{R(z+1)}}{t'_{Rz}} = const. \qquad \text{eq.9.19}$$

Unfortunately the commercial listings of the McReynolds constants no longer include these two values.

---

* McReynolds used the net retention times (t_N: see eq.6.17) in the calculation, which are equal to the adjusted retention times multiplied by the gas compression correction factor (j), the value of which is a constant within one set of measurements. McReynolds wrote eq.2.3 in the following form:

$$\log t_N = a + b \cdot c_n$$

Hence, he called the slope the "b value." It should be noted that the logarithm of r is equal to the slope value (cf. eq.9.18).

9.7 References

[1] E. Kováts, *Helv. Chim. Acta* **41**, 1915–1932 (1958).

[2] P. Toth, E. Kugler and E. Kováts, *Helv. Chim. Acta* **42**, 2519–2530 (1959).

[3] A. Wehrli and E. Kováts, *Helv. Chim. Acta* **42**, 2709–2736 (1959).

[4] L.S. Ettre, *Anal. Chem.* **36** (8), 31A–47A (1964).

[5] E. Kováts, in *Advances in Chromatography, Vol.1* (J.C. Giddings and R.A. Keller, eds.), M. Dekker Inc., New York, 1965; pp. 229–247.

[6] L.S. Ettre, *Chromatographia* **6**, 489–495 (1973).

[7] L.S. Ettre, *Chromatographia* **7**, 39–46 (1974).

[8] L.S. Ettre, *Chromatographia* **7**, 261–268 (1974).

[9] H. van den Dool and P.D. Kratz, *J. Chromatogr.* **11**, 463–471 (1963).

[10] G. Guiochon, *Anal. Chem.* **36**, 661–663 (1964).

[11] J.C. Giddings, in *Gas Chromatography (1961 Lansing Symposium)* (N. Brenner, J.E. Callen and M.D. Weiss, eds.), Academic Press, New York, 1962; pp.57–77.

[12] F.P. Woodford and C.M. van Gent, *J. Lipid Res.* **1**, 188–190 (1960).

[13] T.K. Miwa, K.L. Mikolajczak, F.R. Earle and I.A. Wolff, *Anal. Chem.* **32**, 1739–1742 (1960).

[14] L. Rohrschneider, *J. Chromatogr.* **17**, 1–12 (1966).

[15] L. Rohrschneider, *J. Chromatogr.* **22**, 6–22 (1966).

[16] W.O. McReynolds, *J. Chromatogr. Sci.* **8**, 685–691 (1970),

[17] The data in Table XIX were taken from the catalogs of the supply houses mentioned in the footnote to the table.

PART X

Sample Capacity, Sensitivity and Detection

In this part we shall discuss the relationships concerning the sample capacity of a column and various detector characteristics.

10.1 Sample Capacity

Increasing the sample size above a certain limit creates two problems: It reduces the efficiency of the column and, eventually, overloads the detector, resulting in a non-linear response. Here we deal with the first question; detector overloading is discussed under the linear range of the detector (see Section 10.2).

When investigating the influence of sample capacity, one may consider either the changes in column efficiency, expressed as the number of theoretical plates (plate number) or the HETP, or the changes in peak resolution. Let us not forget that the relationship between plate number and resolution is quadratic (cf. eq.11.8); to double resolution one must increase the plate number by a factor of four.

In their fundamental paper published in 1956 [1], Van Deemter et al. related the sample capacity of a column to the volume of a theoretical plate. According to this, the peak width is not significantly affected by sample size, as long as the concentration of the analyte vapor in the carrier gas plug entering the column is smaller than the concentration of the analyte vapor in the gas volume of one plate. The limiting volume of the analyte vapor (V_{max}) is expressed by eq.10.1:

$$V_{max} = a_K \cdot v_{eff} \sqrt{N} \qquad\qquad \text{eq.10.1}$$

where a_K is a factor, N is the number of theoretical plates, and v_{eff} is the so-called effective volume of one plate:

$$v_{eff} = \frac{V_G + K \cdot V_S}{N} \qquad\qquad \text{eq.10.2}$$

V_G and V_S are the volumes of the gas and stationary (liquid) phases in the column, and K is the partition coefficient. Considering that $K = k \cdot \beta$ and $\beta = V_G/V_S$ (see eqs.4.18b and 4.21; β is the phase ratio of the column and k is the retention factor of the analyte), eq.10.2 can be written as:

$$v_{eff} = \frac{V_G(1+k)}{N} \qquad\qquad \text{eq.10.3}$$

Substituting eq.10.3 into eq.10.1 we obtain:

$$v_{max} = \frac{a_K \cdot V_G(1+k)}{\sqrt{N}} \qquad\qquad \text{eq.10.4}$$

The problem with the general usage of eq.10.4 is that the meaning of factor a_K has never been defined. As stated, it is a semi-empirical quasi-constant that depends somewhat on the analyte and some other, unspecified parameters. Therefore, eq.10.4 cannot be used directly to establish a well-defined sample capacity for a column. However, it had been proposed [2,3] to use the expression on the right-hand side of eq.10.4, without a_K, in comparative evaluations as a value that is *proportional* to the maximum allowed sample size. This term was indicated by the symbol C^*:

$$C^* = \frac{V_G\,(1+k)}{\sqrt{N}}$$ eq.10.5

As discussed in Section 4.3, V_G, the volume of the gas phase in the column, can be calculated from the measured gas holdup volume (V_M) and the compression correction factor (j):

$$V_G = V_M \cdot j$$ eq.4.8b

In the case of open-tubular columns, V_G can also be calculated from the dimensions of the column:

$$V_G = (r_c - d_f)^2\,\pi \cdot L \approx r_c^2 \cdot \pi \cdot L$$ eq.4.7a-b

where L is the column length, r_c is the inner column tube radius and d_f is the liquid phase film thickness.

10.2 Linear Range of a Detector

The *linear range of a chromatographic detector* represents the range of analyte amount or concentration over which the sensitivity of the detector is constant within a specified variation — usually ±5 percent. The linear range is generally presented as the so-called *linearity plot* (see Figure 19, below), plotting detector sensitivity against the injected amount of the analyte, its concentration in the detector, or the mass flow entering it. Since

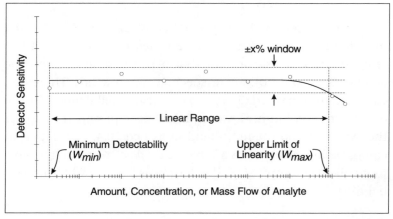

Figure 19. Linearity plot of a chromatographic detector

the linear range of a detector generally extends through a number of orders of magnitude, the linearity plots are usually semi-logarithmic, having a linear scale of the ordinate and a logarithmic scale of the abscissa. The limiting value where the deviation exceeds the specified value (the ±x% window around the plot) can be established from the plot.

The lower limit of linearity is always the *minimum detectable amount* of the analyte, determined separately. Most standard methods specify the minimum detectability as the amount that gives a peak, the height of which is twice the detector noise. However, in practice, one usually considers a factor of five instead of two.

The linear range *(L.R.)* of a detector can also be expressed numerically as the ratio of W_{max} and W_{min} (see Figure 19, page 102):

$$L.R. = W_{max} / W_{min} \qquad\qquad \text{eq.10.6}$$

10.3 Amount (Concentration) of the Analyte in the Detector

Gas chromatographic detectors are generally divided into two groups: concentration-sensitive and mass-sensitive detectors. In the first case we are interested in the concentration of the analyte in the detector, while in the second case, the detector's response depends on the mass flow of the analyte: its amount (mass) entering the detector in unit time.

10.3.1 Concentration of the Analyte in the Detector

The concentration of the analyte in the detector *(c)* is calculated relative to the *peak volume* $(V_p;$ see Section 2.6) and is expressed either in amount/volume or volume/volume units:

$$c = W / V_p \qquad\qquad \text{eq.10.7a}$$

or

$$c = v / V_p \qquad\qquad \text{eq.10.7b}$$

where *W* is the mass (amount) of the analyte present and *v* is the volume of its vapor at column temperature. The value of *v* can be calculated from the mass of the analyte introduced *(W)* and its molecular mass (weight) *(M)*, with the help of the general gas laws:

$$v = \frac{22,414W}{M} \cdot \frac{T_c}{273.15}$$
eq.10.8

where T_c is the absolute column temperature (in degrees Kelvin); *W* and *M* are in grams.

10.3.2 Mass Flow of the Analyte Entering the Detector

The mass (amount) of the analyte entering the detector in unit time (W_t) can be calculated from the mass (amount) corresponding to the peak *(W)* and the peak width at half height (w_h):

$$W_t = W / w_h$$
eq.10.9

where *W* is in mass and w_h in time units; thus the dimension of W_t is mass/time.

10.4 Detector Sensitivity

The *detector sensitivity (S)* is generally expressed as the signal (peak height) for unit concentration or mass (amount) of substance entering the detector in unit time.

In the case of **concentration-sensitive detectors** the sensitivity is:

$$S = m_{max} / c$$
eq.10.10

where m_{max} is the peak height (in mV) and *c* is the concentration of the analyte in the detector (in g/mL). Expressing *c* from eq.10.7a, the detector's sensitivity *(S)* can be written as:

$$S = (m_{max} \cdot V_p) / W$$
eq.10.11

where V_p is the peak volume (see Section 2.6) and *W* is the mass (amount) of the analyte. The dimension of *S* is usually expressed in (mV·mL)/mg.

For thermal-conductivity detectors the sensitivity value is also called the *Dimbat-Porter-Stross (D-P-S) sensitivity* of the detector, after the names of the three scientists who first described it [4].

In the case of **mass-flow sensitive detectors**, the sensitivity is:

$$S = m_{max} / W_t \qquad \text{eq.10.12}$$

where m_{max} is the peak height (in amperes) and W_t is the mass (amount) of the analyte entering the detector in unit time (in g/s). Expressing W_t from eq.10.8, the detector sensitivity can be written as:

$$S = (m_{max} \cdot w_h) / W \qquad \text{eq.10.13}$$

The dimension of S is usually expressed in (amp·s)/g or coulomb/g.

It should be noted that the numerator of both eqs.10.11 and 10.13 is actually the peak area, only the dimensions are different. In eq.10.11 peak area is expressed as mV·mL; in other words, the peak width is given in *volume units*. On the other hand, in eq.10.13 peak area is expressed as amp·s; in this case, the peak width is given in *time units*.

10.5 Detector Selectivity

Detector selectivity is a term used in the case of detectors selective to certain compounds or compound groups. Its value is the ratio of the detector signal obtained for the compound to which the detector has special selectivity vs. the detector signal obtained for a typical compound to which the detector is not particularly sensitive; the signals of both the compound of interest and the reference compound must be obtained or calculated for equal amounts.

10.6 Detector Response Factor

The *detector response factor (f_i)* is a relative term expressing the sensitivity of a detector to a given compound, relative to its sensitivity to a standard compound. As seen in Section 10.4, detector sensitivity *(S)* can be expressed as area *(A)* per concentration or mass (amount) (W). For a given detector, under given conditions, we can write for two compounds that:

$$S_1 = A_1 / W_1 \quad \text{or} \quad W_1 = A_1 / S_1$$

$$S_2 = A_2 / W_2 \quad \text{or} \quad W_2 = A_2 / S_2$$

If $W_1 = W_2$, then $A_1 / S_1 = A_2 / S_2$, and

$$S_2 = S_1 (A_2 / A_1)$$

Let us assume that compound No. 1 is the standard (reference compound: subscript st) and compound No. 2 is the compound of interest (subscript i). If we now use the symbol f instead of S to express the *relative detector response factors,* we can write that:

$$f_i = f_{st} (A_i / A_{st}) \qquad\qquad \text{eq.10.14}$$

The relative detector response factors can be expressed either in relative molar or in equal mass (weight) basis. In the case of *relative molar responses,* using n-alkanes as the standards, the assigned value of f_{st} is usually the number of carbon atoms of the n-alkane multiplied by 100 (e.g., 600 for n-hexane). If the relative detector response is expressed on an equal mass (weight) basis, then usually an arbitrary value of 1 or 100 is assigned to f_{st}.

10.7 References

[1] J.J. Van Deemter, F.J. Zuiderweg and A. Klinkenberg, *Chem. Eng. Sci.* **5**, 271–289 (1956).

[2] L.S. Ettre, *Chromatographia* **18**, 477 (1984).

[3] L.S. Ettre, in *The Science of Chromatography* (F. Bruner, ed.), Elsevier, Amsterdam, 1985; pp. 87–109.

[4] M. Dimbat, P.E. Porter and F.H. Stross, *Anal. Chem.* **28**, 290–297 (1956).

PART XI

Fundamental Relationships of Chromatography

There are a number of fundamental relationships in (gas) chromatography expressing the relationship between the characteristics of analytes, column dimensions and other important chromatographic parameters such as retention, efficiency, separation and resolution. Understanding these relationships helps one properly adjust the column parameters and the analytical conditions to an actual problem.

11.1 Partition Coefficient vs. Retention Factor and Phase Ratio

We have already described this relationship in Sections 4.5 and 4.8. It relates the partition coefficient (distribution constant) (K) to the retention factor (capacity ratio) (k) and the phase ratio of the column (β). In the case of open-tubular columns the relationship includes the inner tube radius (r_c) or the tube diameter (d_c) and the stationary phase film thickness (d_f):

$$K = k \cdot \beta \qquad \text{(packed columns)}$$
$$\text{eq.11.1a}$$

$$K = k \cdot \beta = \frac{k \cdot r_c}{2 d_f} = \frac{k \cdot d_c}{4 d_f} \qquad \text{(open-tubular columns)}$$
$$\text{eq.11.1b}$$

This relationship represents the basis for the variation of the liquid phase loading of packed columns or the variation of the diameter and film thickness of open-tubular columns. For example, for a high-boiling analyte that elutes late, the retention factor (k) can be reduced by increasing the phase ratio (β), because at a given temperature their product is constant. In the case of packed columns, the phase ratio can be increased by decreasing the liquid phase loading, while in the case of open-tubular columns this can be achieved by increasing the tube radius or reducing the stationary phase film thickness — or both. Conversely, in the case of low-boiling analytes, it is advantageous to increase their retention factors (see Section 11.3); this can be achieved by increasing the liquid phase loading or the stationary phase film thickness.

Similar changes can also be achieved by changing the column temperature: increasing it when high-boiling compounds are analyzed and reducing it for the analysis of low-boiling compounds.

11.2 Retention Volume vs. Phase Ratio and Partition Coefficient

The retention factor k (cf. eq.3.1) can also be written using retention volumes instead of times:

$$k = \frac{t_R - t_M}{t_M} = \frac{V_R - V_M}{V_M} \qquad \text{eq.11.2a}$$

If we multiply both the numerator and the denominator by the gas compression correction factor (j), we shall have the corrected retention volume of the analyte (V_R^o) and the corrected gas holdup volume (V_M^o) in the right-hand side fraction of the equation:

$$k = \frac{(V_R - V_M)j}{V_M \cdot j} = \frac{V_R^o - V_M^o}{V_M^o} \qquad \text{eq.11.2b}$$

We have seen earlier (Section 4.8) that the phase ratio (β) is equal to the ratio of the volumes of the gas $(V_G = V_M^o)$ and stationary (V_S) phases in the column (cf. eq.4.18b):

$$\beta = V_G/V_S = V_M^o/V_S$$

Substituting this expression for β and the right-hand side of eq.11.2b for k into eq.11.1a, we obtain the following relationship:

$$K = \beta \cdot k = \frac{V_M^o}{V_S} \cdot \frac{V_R^o - V_M^o}{V_M^o} = \frac{V_R^o - V_M^o}{V_S} \qquad \text{eq.11.3}$$

Expressing V_R^o from eq.11.3:

$$V_R^o = V_M^o + K \cdot V_S \qquad \text{eq.11.4a}$$

or

$$V_R^o = V_G + K \cdot V_S \qquad \text{eq.11.4b}$$

This is a fundamental relationship in gas chromatography, indicating how the partition coefficient (distribution constant) and the volumes of the two phases present in the column influence retention.

We may carry out one more modification of eq.11.3 by substituting the net retention volume (V_N) of the analyte for $(V_R^o - V_M^o)$ and using the amount (weight) of the stationary phase (W_S) instead of its volume; ρ_S is the density of the stationary phase at column temperature:

$$V_S = W_S/\rho_S \qquad \text{eq.11.5}$$

$$K = \frac{V_N \cdot \rho_S}{W_S} \qquad \text{eq.11.6}$$

It should be noted that for liquid chromatography, where mobile phase compression can be neglected (hence, $j = 1.00$), eq 11.4a can be written as:

$$V_R = V_M + K \cdot V_S \qquad \text{eq.11.7}$$

where V_M is now the mobile phase volume in the column, which is also equal to the holdup volume.

11.3 Efficiency and Resolution vs. Selectivity and Retention

This relationship combines efficiency (plate number), resolution, selectivity (relative retention) and retention (retention factor), and expresses their interrelationship. It is based on the relationships describing the plate number (N; eq.7.2a); resolution (R_s; eq.8.10); separation factor (α; eq.8.1) and retention factor (k; eq.3.1) as a function of the total (t_R) and adjusted (t'_R) retention times, the holdup time (t_M); and the peak width a base (w_h):

$$N = 16(t_R / w_b)^2 \qquad\qquad \text{eq.7.2a}$$

$$R_s = \frac{2(t'_{R2} - t'_{R1})}{w_{b1} + w_{b2}} \qquad\qquad \text{eq.8.10}$$

$$\alpha = t'_{R2}/t'_{R1} \qquad\qquad \text{eq.8.1}$$

$$k = t'_R/t_M \qquad\qquad \text{eq.3.1}$$

Depending on the interpretation of eq.7.2a and the selection of the peak from the peak pair to be separated, this fundamental relationship exists in three versions. They are derived in detail in Supplement No.III; here we only summarize the concept and give the final equations.

We should note here that the final equations can be written in two ways: either expressing the plate number or the resolution. In the first case, we obtain the *number of theoretical plates required (N_{req})* to separate with resolution R_s two peaks having a separation factor α and the retention factor k of the specified peak. In the second case, we calculate the resolution (R_s) of the specified two peaks that can be achieved on a certain column having a specified plate number (N). In addition, we can also substitute L/H (cf. eq.7.3) for the plate number and thus express resolution in terms of the column length (L) and plate height (H).

11.3.1 First Version

This relationship, first described by Purnell [1], is based on the assumption that $w_{b1} = w_{b2}$ and considers the **second** peak of the peak pair. Thus, it utilizes eq.8.11 to express resolution:

$$R_s = (t'_{R2} - t'_{R1}) / w_{b2} \qquad \text{eq.8.11}$$

Consequently, the plate number is also expressed for the second peak. The final relationships are:

$$\frac{L}{H_2} = N_{req} = 16R_s^2 \left(\frac{\alpha}{\alpha-1}\right)^2 \left(\frac{k_2+1}{k_2}\right)^2 \qquad \text{eq.11.8}$$

$$R_s = \frac{\sqrt{N_2}}{4} \cdot \frac{\alpha-1}{\alpha} \cdot \frac{k_2}{k_2+1} = \frac{1}{4}\sqrt{\frac{L}{H_2}} \cdot \frac{\alpha-1}{\alpha} \cdot \frac{k_2}{k_2+1} \qquad \text{eq.11.9}$$

11.3.2 Second Version

This relationship was first described by Knox [2] and Thijssen [3]. It differs from the first version in the selection of the peak considered: It is now the **first** peak. Therefore, eq.8.11 is now written as:

$$R_s = (t'_{R2} - t'_{R1}) / w_{b1}$$

and the plate number is also expressed for the first peak. The final relationships are:

$$\frac{L}{H_1} = N_{req} = 16R_s^2 \left(\frac{1}{\alpha-1}\right)^2 \left(\frac{k_1+1}{k_1}\right)^2 \qquad \text{eq.11.10}$$

$$R_s = \frac{\sqrt{N_1}}{4} \cdot (\alpha-1) \cdot \frac{k_1}{k_1+1} = \frac{1}{4}\sqrt{\frac{L}{H_1}} \cdot (\alpha-1) \cdot \frac{k_1}{k_1+1} \qquad \text{eq.11.11}$$

11.3.3 Third Version

This relationship was first described by Said[4]. He considered a hypothetical solute characterized by the average total retention time ($\bar{t}_R$), the average peak width at base ($\bar{w}_b$) and the average retention factor ($\bar{k}$), and calculated the value of the theoretical plate number N^* from these values:

$$\bar{t}_R = \frac{t_{R1} + t_{R2}}{2} \qquad \text{eq.11.12}$$

$$\overline{w}_b = \frac{w_{b1} + w_{b2}}{2} \qquad \text{eq.11.13}$$

$$\overline{k} = \frac{k_1 + k_2}{2} \qquad \text{eq.11.14}$$

$$N^* = 16(\overline{t}_R/\overline{w}_b)^2 \qquad \text{eq.11.15}$$

Note that the plate number N^* calculated in this way is equal neither to the individual plate numbers nor to the mean value of the individual plate numbers:

$$N^* \neq N_1 \neq N_2 \neq \overline{N}$$

where

$$\overline{N} = \frac{N_1 + N_2}{2}$$

The final relationships of the third version are (H^* is the HETP value corresponding to N^*):

$$\frac{L}{H^*} = N^*_{req} = 4R_s^2 \left(\frac{\alpha+1}{\alpha-1}\right)^2 \left(\frac{\overline{k}+1}{\overline{k}}\right)^2 \qquad \text{eq.11.16}$$

$$R_s = \frac{\sqrt{N^*}}{2} \cdot \frac{\alpha-1}{\alpha+1} \cdot \frac{\overline{k}}{\overline{k}+1} = \frac{1}{2}\sqrt{\frac{L}{H^*}} \cdot \frac{\alpha-1}{\alpha+1} \cdot \frac{\overline{k}}{\overline{k}+1} \qquad \text{eq.11.17}$$

11.3.4 Comparative Values

The values calculated according to the different versions actually differ very little from each other. Taking as an example a typical case with $\alpha = 1.05$ and two retention factors of $k_1 = 4.76$ and $k_2 = 5.00$, and assuming a resolution of $R_s = 1.5$, the following results are obtained for the required number of theoretical plates:

First version (eq.11.8): $N_{req} = 22,861$

Second version (eq.11.10): $N_{req} = 21,086$

Third version (eq.11.16): $N_{req} = 21,965$

The relationships discussed above provide an excellent way to assess the influence of the separation factor (selectivity) and the retention factor on column efficiency needed for a desired resolution or on the resolution that can be obtained for a given column (fixed efficiency). Table XX, below, lists values of N_{req} as a function of α and k. For example, if $\alpha = 1.05$ and baseline resolution $(R_s = 1.5)$ is desired, and if the peaks are located in the very early part of the chromatogram $(k_2 = 0.1)$, then we would need close to two million theoretical plates for the separation. On the other hand, if we can adjust the conditions, e.g., by using a thicker-film column or a lower column temperature (cf. Section 11.1), then the value of the retention factor will be increased and the required number of theoretical plates will rapidly decrease.

11.3.5 Use of the Unadjusted Relative Retention

The separation factor (α) used in the relationships discussed above is the ratio of the adjusted retention times (t'_R) or retention factors (k) of two consecutive peaks:

$$\alpha = \frac{t'_{R2}}{t'_{R1}} = \frac{k_2}{k_1} \qquad\qquad eq.8.1$$

α	k_2	N_{req}
	0.1	1,920,996
1.05	0.5	142,884
	1.0	63,504
	5.0	22,861
	0.1	527,076
1.10	0.5	39,204
	1.0	17,424
	5.0	6,273

Table XX. Values of N_{req} *for different values of* α *and* k, *calculated from eq.11.8* ($R_s = 1.5$).

However, as we have seen earlier (see Section 8.2), a similar value can also be calculated by using the total retention times (t_R). This term is called the *unadjusted separation factor* (α_G):

$$\alpha_G = \frac{t_{R2}}{t_{R1}} = \frac{k_2 + 1}{k_1 + 1} \qquad \text{eq.8.7a}$$

If we would use α_G instead of α in the relationships discussed in Sections 11.3.1–11.3.3, then the following final relationships would be obtained for the most frequently used first version:

$$\frac{L}{H_2} = N_{req} = 16R_s^2 \left(\frac{\alpha_G}{\alpha_G - 1} \right)^2 \qquad \text{eq.11.18}$$

$$R_s = \frac{\sqrt{N_2}}{4} \cdot \frac{\alpha_G - 1}{\alpha_G} = \frac{1}{4}\sqrt{\frac{L}{H_2}} \cdot \frac{\alpha_G - 1}{\alpha_G} \qquad \text{eq.11.19}$$

Similar relationships can also be derived for the other two versions.

11.4 Van Deemter-Golay Equations

The fundamental equations describing the contributions of various processes in the gas chromatographic column were originally described in 1956–1958 in two basic papers by Van Deemter, Zuiderweg and Klinkenberg [5] for packed columns and by Golay [6] for open-tubular columns. In the decades following these original publications a number of papers have dealt with various aspects of these relationships, and every chromatography textbook has separate chapters discussing in detail the influence of the individual parameters included in these equations on column efficiency. Here we will only summarize the individual terms of these relationships that are valid for gas chromatography.

Both the so-called Van Deemter and Golay equations express the HETP of a column *(H)* as a function of the average linear mobile phase velocity $(\overline{u})$ and of three processes taking place in the column during the passage of the analyte. The basic form of these relationships is:

$$H = A + B/\overline{u} + C \cdot \overline{u} \qquad\qquad eq.11.18$$

The three terms represent the effects of the multiplicity of gas paths ("eddy diffusion") in the column *(A)*, longitudinal gas-gas diffusion in the mobile phase *(B)*, and the so-called resistance to mass transfer in the column related to the diffusion process in the mobile (C_M) and stationary (C_S) phases:

$$C = C_M + C_S \qquad\qquad eq.11.19$$

In open-tubular columns where the gas path is unobstructed, the *A* term is equal to zero and eq.11.18 is reduced to the Golay equation:

$$H = B/\overline{u} + C \cdot \overline{u} \qquad\qquad eq.11.20$$

The relationships described in eqs.11.18 and 11.20 correspond to the hyperboles shown in Figure 20 on page 116; the figures also indicate the plots corresponding to the individual terms in the equations. The minima of the plots, at optimum average carrier gas velocity $(\overline{u}_{opt})$, correspond to the smallest value of HETP $(HETP_{min}, H_{min})$, i.e., to the best column performance. Usually, however, one does not operate a column at optimum but at a somewhat higher velocity. If the carrier gas velocity is set at optimum for one test compound, then during a chromatographic run one may inadvertently operate the column below the optimum velocity for other peaks. On the other hand, below optimum, the *B* term (longitudinal diffusion) will dominate the process in the column, and this is detrimental for separation (cf. Figure 20, page 116). There are two reasons why this can happen. The first is connected with programmed-temperature operation in GC systems where the inlet pressure (i.e., the pressure drop, Δp) is kept constant. Since the carrier gas viscosity (η) increases with increasing temperature, the actual average linear carrier gas velocity $(\overline{u}$) will decrease in order to maintain a constant pressure drop (see eq.5.23):

$$\Delta p \cdot j' = \frac{8L \cdot \eta \cdot \overline{u}}{r_c^2} \qquad\qquad eq.5.23$$

Thus, if the velocity is set at a lower temperature, then the

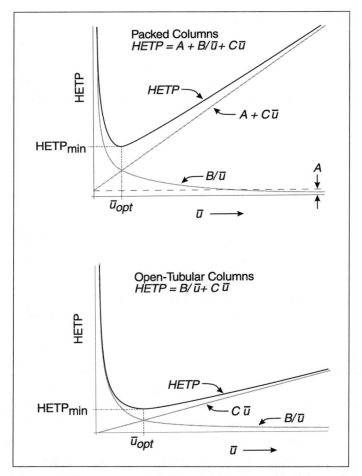

Figure 20. Plots of HETP vs. $\bar{u}$ for packed and open-tubular columns

average velocity will be reduced during the program. This change might be significant: For example, programming from 100°C to 300°C increases the viscosity of helium by 35%, causing a corresponding reduction in the average linear gas velocity.

The second reason is due to the fact that one can optimize the velocity for only one component of the sample: If we worked at optimum for this one, then for some other sample components this velocity would be below their optimum value.

Because we usually work above the optimum velocity, the slope of the HETP vs. $\bar{u}$ plot is an important characteristic: the smaller the slope the less will be the increase in HETP upon increasing $\bar{u}$. In general, we can state that the slope is smaller for open-tubular columns than for packed columns.

Looking at eqs.11.18 and 11.20, it is clear that the ascending part of the plots after $\bar{u}_{opt}$ approach the relationship:

$$H \approx C \cdot \bar{u} \qquad \text{eq.11.21}$$

In other words, the C term of the equations represents the slope of the ascending part of the plots.

Another important characteristic of these plots is the location of $\bar{u}_{opt}$: It is advantageous to have it at a higher velocity because this reduces the time of analysis. In general, the value at optimum velocity is smaller for a packed column than for an open-tubular column.

Both the slope of the plots and the value of the optimum velocity also depend on the carrier gas. For a given column, the slope will be higher and $\bar{u}_{opt}$ smaller for high-density gases (such as nitrogen) than for low-density gases (such as helium and, particularly, hydrogen).

11.4.1 Packed Columns (Van Deemter Equation)

A number of forms can be found in the literature for the four terms A, B, C_M and C_S in eqs.11.18–11.19. Most of these contain constants that are not well defined, mainly because of differences in the geometry of the column packing. The most common forms of these terms for packed columns are:

$$A = 2\lambda \cdot d_p \qquad \text{eq.11.22}$$

$$B = 2\gamma \cdot D_M \qquad \text{eq.11.23}$$

$$C_M = \frac{\omega \cdot d_p^2}{D_M} \qquad \text{eq.11.24}$$

$$C_S = \frac{q \cdot k}{(k+1)^2} \cdot \frac{d_f^2}{D_S} \qquad \text{eq.11.25}$$

where d_p is the particle diameter, d_f is the thickness of the stationary (liquid) phase coating, k is the retention factor (capacity ratio), and D_M and D_S are the diffusion coefficients of the analyte in the mobile and stationary phases, respectively. The four constants are identified in the literature as follows:

λ   is the so-called *packing factor,* with a value usually between 1 and 2;

γ   is the so-called *tortuosity factor,* a quasi-constant that is not totally independent of mobile phase velocity, with a value about 0.6–0.8;

ω   is another *packing factor* intended to correct for radial diffusion, with a value between 0.02 and 5;

q   is the so-called *configuration factor;* in the case of a uniform film coated on a solid surface, its value is 2/3.

11.4.2 Open-Tubular Columns (Golay Equation)

Because of the uniformity of the flow path and coating, the individual terms of the plate height equation (eq.11.18) are rigorously calculable in the case of open-tubular columns:

- because there is no multiplicity of the mobile phase flow, the *A* term is eliminated;
- because there are no obstacles in the flow path, the tortuosity factor *(γ)* in the *B* term is equal to unity;
- function *ω* in the C_M term is now expressed as:

$$\omega = \frac{1+6k+11k^2}{96(1+k)^2} \qquad \text{eq.11.26}$$

and the inner diameter *(d_p)* is substituted for the particle diameter;

- As previously mentioned, the value of the configuration factor *(q)* in the C_S term is 2/3 in the case of a uniformly coated film.

As a conclusion of these considerations the three terms of the Golay equation:

$$H = \frac{B}{\bar{u}} + \left(C_M + C_S\right)\cdot\bar{u} \qquad \text{eq.11.27a}$$

can be written in the following form:

$$B = 2D_M \qquad \text{eq.11.28}$$

$$C_M = \frac{1+6k+11k^2}{96(1+k)^2} \cdot \frac{d_c^2}{D_M} = \frac{1+6k+11k^2}{24(1+k)^2} \cdot \frac{r_c^2}{D_M} \qquad \text{eq.11.29}$$

$$C_S = \frac{2k}{3(1+k)^2} \cdot \frac{d_f^2}{D_S} \qquad \text{eq.11.30a}$$

We may express d_f from eq.4.13 as $d_f = r_c \cdot k / 2K$. Substituting this into eq.11.30a, it can be written as:

$$C_S = \frac{k^3}{6(1+k)^2} \cdot \frac{r_c^2}{K^2 \cdot D_S} = \frac{k^3}{24(1+k)^2} \cdot \frac{d_c^2}{K^2 \cdot D_S} \qquad \text{eq.11.30b}$$

It should be noted that in Golay's original paper [6] eq.11.30a was written in a different form, with an additional term in the denominator:

$$C_S = \frac{2k}{3(1+k)^2} \cdot \frac{d_f^2}{F^2 \cdot D_S} \qquad \text{eq.11.31}$$

Here F represented the ratio of the inside tube area coated with the liquid (stationary) phase (A_{coated}) to the geometric inside area of a smooth-walled tube with a radius r_c and a length of L:

$$F = \frac{A_{coated}}{(2r_c \pi) L} \qquad \text{eq.11.32}$$

By adding this term to the expression of resistance to mass transfer in the stationary phase, Golay foresaw the possibility of improvement by increasing the inside tube surface area to be coated, and thus reducing the film thickness without reducing the total volume of the stationary phase in the column. In subsequent works, because in general the inside tube surface was assumed to be smooth, the F term was neglected (or, more correctly, assumed to be equal to unity). However, this assumption is not always valid. In the case of a roughened — etched — inside surface, one should definitely consider this term. Note particularly that we have the square of F in the denominator, indicating the particularly beneficial influence of an increase of the coated surface.

It should be mentioned that in the case of truly thin-film

columns, the C_S term of eq.11.27a may be neglected and the relationship written in the following simplified form:

$$H = \frac{B}{\bar{u}} + C_M \cdot \bar{u}$$

eq.11.27b

However, in the case of thick-film columns this simplification is no longer valid: there the influence of the C_S term is significant. Note particularly that eq.11.30a contains the square of the film thickness; thus, doubling the film thickness will increase the value of the C_S term by a factor of four.*

Table XXI on page 121 lists values of certain functions of k present in eqs.11.30a–b, 11.35a–b, 11.36a–b and 11.39.

11.4.3 H and $\bar{u}$ Values at Optimum

Differentiating eq.11.18 with respect to $\bar{u}$ and setting the result equal to zero gives expressions for the optimum value of mobile phase velocity ($\bar{u}_{opt}$) and the corresponding minimum of the plate height ($HETP_{min}$):

$$\bar{u}_{opt} = \sqrt{B/(C_M + C_S)}$$

eq.11.33a

$$HETP_{min} = A + 2\sqrt{B(C_M + C_S)}$$

eq.11.33b

In the case of **open-tubular columns**, these relationships can be further simplified. First, $A = 0$; in addition, we assume that resistance to mass transfer in the stationary phase (i.e., the C_S term) can be neglected. Thus, the optimum values can be written as:

$$\bar{u}_{opt} = \sqrt{B/C_M}$$

eq.11.34a

$$HETP_{min} = 2\sqrt{B \cdot C_M}$$

eq.11.34b

---

* Strictly speaking, this is true only if we disregard the first fraction on the right-hand side of eq.11.30a. Increasing the film thickness will also correspondingly increase the value of the retention factor (k), and this will change the value of this fraction, which, after going through a maximum at $k = 1$, will be reduced with increasing values of k (see Table XXI, page 121).

k	f_1	f_2	f_3	f_4	f_5	f_6
0.0	0.0000	∞	6.9282	13.8564	0.5774	0.2887
0.1	0.0551	133.1000	5.8279	11.6559	0.6863	0.3432
0.5	0.1481	13.5000	4.0000	8.0000	1.0000	0.5000
1.0	0.1667	8.0000	3.2660	6.5320	1.2247	0.6124
1.5	0.1600	6.9444	2.9382	5.8764	1.3614	0.6807
2.0	0.1481	6.7500	2.7530	5.5060	1.4530	0.7265
2.5	0.1361	6.8600	2.6340	5.2680	1.5186	0.7593
5.0	0.0926	8.6400	2.3764	4.7527	1.6833	0.8416
10.0	0.0551	13.3100	2.2367	4.4733	1.7884	0.8942
15.0	0.0391	18.2044	2.1883	4.3767	1.8279	0.9139
20.0	0.0302	23.1525	2.1638	4.3277	1.8486	0.9243
25.0	0.0247	28.1216	2.1490	4.2980	1.8613	0.9307
50.0	0.0128	53.0604	2.1191	4.2383	1.8876	0.9438
100.0	0.0065	103.0301	2.1041	4.2082	1.9011	0.9505

$$f_1 = \frac{2k}{3(1+k)^2} \qquad \text{eq.11.30a}$$

$$f_2 = \frac{(1+k)^3}{k^2} \qquad \text{eq.11.44}$$

$$f_3 = \sqrt{\frac{48(1+k)^2}{1+6k+11k^2}} \qquad \text{eq.11.35a}$$

$$f_4 = \sqrt{\frac{192(1+k)^2}{1+6k+11k^2}} \qquad \text{eq.11.35b}$$

$$f_5 = \sqrt{\frac{1+6k+11k^2}{3(1+k)^2}} \qquad \text{eq.11.36a}$$

$$f_6 = \sqrt{\frac{1+6k+11k^2}{12(1+k)^2}} \qquad \text{eq.11.36b}$$

Table XXI. Values of certain functions of k

Substituting eq.11.28 for B and eq.11.29 for C_M we obtain:

$$\bar{u}_{opt(theor)} = \frac{D_M}{r_c}\sqrt{\frac{48(1+k)^2}{1+6k+11k^2}}$$ eq.11.35a

$$HETP_{min(theor)} = r_c\sqrt{\frac{1+6k+11k^2}{3(1+k)^2}}$$ eq.1136a

where k is the retention factor (capacity ratio). Both equations can also be written using the inner tube diameter (d_c) instead of the radius (r_c):

$$\bar{u}_{opt(theor)} = \frac{D_M}{d_c}\sqrt{\frac{192(1+k)^2}{1+6k+11k^2}}$$ eq.11.35b

$$HETP_{min(theor)} = d_c\sqrt{\frac{1+6k+11k^2}{12(1+k)^2}}$$ eq.11.36b

Investigating the relationship describing the **theoretically obtainable minimum HETP** ($HETP_{min(theor)}$), we can see that its value is directly related to the inner tube diameter (radius). Thus, for high efficiency, one needs small diameter columns.

Table XXI on page 121 presents values for the square root function; in eq.11.36b its value varies between 0.29 for $k = 0$ and 0.95 for $k = 100$. This means that the theoretically obtainable best HETP varies from 29% of the tube diameter to about the tube diameter. It also means that the HETP of early peaks is smaller than for later peaks. We should mention here that this increase in column efficiency will not compensate for the need for even higher efficiencies at early peaks (see Section 11.3.4).

Eq.11.36b does not include the gaseous diffusion coefficient, and thus the value of the calculated theoretically best HETP ($HETP_{min(theor)}$) is independent of the carrier gas used. Actually, the carrier gas has a small influence on its value. This can be seen in Figure 21 on page 123, which presents typical plots for a thin-film open-tubular column $(d_c = 0.25\ mm)$ for three carrier gases, at 175°C, using n-heptadecane $(k = 4.95)$ as the

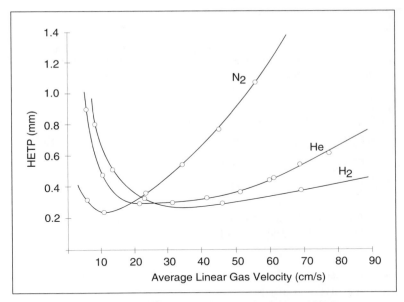

Figure 21. HETP vs. $\overline{u}$ plots for n-heptadecane at 175°C on an open-tubular column coated with OV-101 methylsilicone phase, for three carrier gases. Column dimensions: 25 m × 0.25-mm I.D. Liquid phase film thickness: 0.4 μm [7].

analyte [7]; Table XXII on page 124 lists the values at optimum conditions, established from these plots and the corresponding theoretical values. As seen, there is a difference of about 20% between the lowest value (measured with nitrogen) and the highest value (measured for helium). Since, however, the slope of the ascending part of the plots for helium and hydrogen is much smaller than for nitrogen, these differences can be compensated by using a slightly longer column at a correspondingly higher velocity.

Concerning the agreement between the theoretically calculated and actually measured $HETP_{min}$ values, our practice has shown that the latter are larger than the theoretical values by a factor of about 1.25–1.6; the agreement is better for smaller-diameter and thin-film columns. The reason for this difference is essentially twofold. As seen (cf. eq.11.34b vs. eq.11.33b), the C_S term was neglected; this omission is really unjustified for thick-film columns where C_S may actually be the dominant term. The

		Carrier gas		
		Nitrogen	Helium	Hydrogen
D_M for n-heptadecane	cm²/s	0.093	0.306	0.383
$\overline{u}_{opt\,(theor)}$	cm/s	17.7	58.2	72.9
$\overline{u}_{opt}$ measured	cm/s	12	22	36
$HETP_{min(theor)}$	mm	0.21	0.21	0.21
$HETP_{min}$ measured	mm	0.23	0.29	0.27
U.T.E.	%	91.3	72.4	77.8

*Table XXII. Typical values corresponding to the plots in Figure 21 on page 123, together with the calculated theoretical values. ***

other reason for the difference may be unevenness in the stationary phase coating.

The difference between the calculated $HETP_{min(theor)}$ and the measured value may be used as a column characteristic; see Section 11.4.4.

Concerning the relationship describing the **theoretical optimum average linear gas velocity** ($\overline{u}_{opt\,(theor)}$), the velocity corresponding to the minimum of the Van Deemter-Golay plots, a number of important conclusions can be drawn.

As shown by eq.11.35b, its value depends on three terms: the inner tube diameter (d_c); the diffusion coefficient of the analyte in the mobile phase (D_M), which is a function of the compound, the temperature and the carrier gas; and the square root term including the retention factor (k). The optimum velocity is an inverse function of the tube diameter; this means that in the case of larger-diameter columns the optimum velocity will be less (disregarding at this moment the possible change in the retention factor, influencing the square root term).

---

* The D_M values were calculated using the Fuller-Schettler-Giddings equation (eq.5.27), assuming a pressure of 1 atm. The $\overline{u}_{opt\,(theor)}$ and $HETP_{min(theor)}$ values were calculated using eqs.11.35b and 11.36b, respectively, while the utilization of theoretical efficiency (U.T.E. %) values were calculated using eq.11.37. The measured values were established from the plots shown in Figure 21.

It is important to investigate the influence of the diffusion coefficient (D_M). Three factors influence the value of D_M, and thus $\overline{u}_{opt}$. The first concerns changes due to the compound. Table X on page 47 lists values for n-paraffins. As seen, the diffusion coefficient decreases with molecular weight. Since in general we may state that within one sample the higher molecular weight compounds emerge later, this means that the optimum velocity for later peaks will be less than that for earlier peaks, and this change is enhanced even more by the fact that the value of the square root term also decreases with increasing values of k (see Table XXI, page 121). For any peak, we should not use a velocity less than the optimum. This means that when establishing the actual velocity for a chromatographic run, we should not use an early peak for optimization and we should definitely select a velocity that is higher than the optimum for the selected test substance.

The second factor influencing the value of D_M and, hence, of $\overline{u}_{opt(theor)}$, is temperature. According to the Fuller-Schettler-Giddings equation (eq.5.27), D_M is proportional to $(T_c)^{1.75}$ (where column temperature is in degrees Kelvin). This means that at a higher temperature the value of D_M for a given analyte will be higher. This is somewhat (but not completely) compensated for by the fact that at higher temperatures, the value of the square root term is reduced (cf. Table XXI, page 121). In other words, for the same analyte, the optimum velocity will have a higher value at a higher temperature.

The third factor influencing the value of D_M is the carrier gas. Table X on page 47 gave values for n-alkanes for three carrier gases; it was the highest for hydrogen (the lowest-density gas) and the lowest for nitrogen (the highest-density gas). This means that the highest optimum velocity is obtained when using hydrogen as the carrier gas. The difference is significant: For example, for n-decane the value of D_M for hydrogen is 23% higher than for helium and higher by a factor of four than the value for nitrogen.

Finally, we should investigate the square root term in eq.11.35b. As shown in Table XXI, its value changes from about 14 (at $k = 0$) to 4.2 (at $k = 100$). At low k values the change is significant; however, after about $k = 2$, the difference is relatively minor: Between $k = 2$ and $k = 5$ its value is reduced by only 13.7%.

Concerning the agreement between the theoretically calculated and actually measured optimum velocity values, we may compare the values corresponding to Figure 21 on page 123 (Table XXII, page 124). The interesting conclusion of the evaluation of these data is that, while in the case of actual vs. theoretical $HETP_{min}$ the actual values are larger by a factor of 1.1 to 1.3, the difference in the case of $\overline{u}_{opt}$ is much larger: here the theoretical values are larger by a factor of 1.5 to 2.0. In addition to the two reasons already mentioned for $HETP_{min}$, this larger discrepancy may also be connected to the way D_M is usually calculated by using the outlet pressure (1 atm) for p in eq.5.27. In fact, the actual D_M that would correspond to $\overline{u}_{opt\,(theor)}$ would be smaller than the calculated value because D_M is inversely proportional to pressure and the pressure in the column is higher than the outlet pressure. In other words, $\overline{u}_{opt\,(theor)}$ calculated using eqs.11.35a–b and considering D_M at outlet pressure will be larger than the "real" $\overline{u}_{opt\,(theor)}$. In addition we should not forget that the Fuller-Schettler-Giddings equation used for the calculation of D_M is not an exact but only an empirical relationship.

11.4.4 Utilization of the Theoretical Efficiency

This term is widely used as the basis of comparison of actual column efficiency to the theoretically obtainable best performance ($HETP_{min(theor)}$: eqs.11.36a–11.36b). In this comparison we calculate $HETP_{min(theor)}$ for a given column and analyte (i.e., for a given d_c and k). In contrast, the *measured* plate height is always larger than $HETP_{min(theor)}$. Expressing this relationship as a percent value gives us the *utilization of the theoretical efficiency (U.T.E.%)*:

$$U.T.E.\% = \frac{HETP_{min}(theoretical)}{HETP(measured)}100 \qquad \text{eq.11.37}$$

This term was sometimes also called the "coating efficiency," implying that it indicates how efficient the coating procedure was. However, this is an incorrect interpretation of its meaning.

For small-diameter, thin-film open-tubular columns, the value of U.T.E.% should be over 70% (see also Table XXII, page 124). For larger diameter and thick-film columns, its value is around 60%. The main reason for this poorer value is that in such a case the C_S term cannot be neglected anymore. However, the U.T.E.% is a very useful expression even in such a case because it compares actual column performance of a thick-film column to that of an idealized, thin-film column with the same diameter.

The concept of the utilization of the theoretical column efficiency was first elaborated by Desty et al. [8]. The term was first used by Ettre [9].

11.4.5 The Extended Golay Equation

As mentioned above, in his original treatment Golay had already considered the possibility of increasing the inner tube area to be coated by including F^2 in the denominator of the C_S term (cf. eq.11.31). Ten years later, after the introduction of PLOT and SCOT columns, he developed a generalized equation to describe the efficiency of open-tubular columns which is valid for all types of columns [10]. In this extended Golay equation the C_M and C_S terms are written in the following way:

$$C_M = \left[\frac{1 + 6k + 11k^2}{(1+k)^2} + \frac{8 + 32k}{(1+k)^2} \cdot a_2 + \frac{8k^2}{(1+k)^2} \cdot \frac{a_1^2}{a_2} \right] \frac{r_c^2}{24 D_G} \qquad \text{eq.11.38}$$

$$C_S = \frac{k^3}{6(1+k)^2} \cdot \frac{1 + 2a_2}{F^2} \cdot \frac{r_c^2}{K^2 \cdot D_S} \qquad \text{eq.11.39}$$

It should be mentioned that in ref. [10], eq.11.38 is printed with an error; in the term at the end of the right-hand side, r_c is given instead of the correct r_c^2.

There are two new terms in eqs.11.38–11.39: a_1 and a_2, which characterize the porous layer on the inner wall of the column tubing. It is assumed that this porous layer represents a tortuous path, the length of which is $a_1 \cdot r_c$, and it is characterized by an average free gas volume, the thickness of which is $a_2 \cdot r_c$. If we describe the thickness of the porous layer (d_{pl}) as a fraction of the tube radius (r_c):

$$d_{pl} = a \cdot r_c \qquad\qquad \text{eq.11.40}$$

then we can express a_1 and a_2 with respect to a:

- a_1 is larger than a;
- a_2 is smaller than a;
- all three values are much smaller than unity:

$$a_1 > a \qquad\qquad\qquad a_2 < a$$

$$a << 1 \qquad\quad a_1 << 1 \qquad\quad a_2 << 1$$

The essential problem with eqs.11.38 and 11.39 is that it is almost impossible to accurately estimate the values of a_1 and a_2. As pointed out by Ettre and Purcell [11], the best one can achieve is to set certain values for a_1, and calculate the corresponding values for a_2.

Comparing eq.11.38 with eq.11.29 and eq.11.39 with eq.11.30b, we can see that if we assume that $a_1 = a_2 = 0$ and $F = 1$, i.e., the smooth (geometric) inner surface of the tube is coated with a film of the stationary phase, then eqs.11.38 and 11.39 will be identical to eqs.11.29 and 11.30b, describing wall-coated open-tubular columns.

11.4.6 The Giddings Equation

The Van Deemter-Golay equations contain the diffusion coefficient in the carrier gas (D_M) in both the B and C_M terms (see eqs.11.23–11.24 and eqs.11.28–11.29). As shown by eq.5.27 (Section 5.6.1), its value is inversely proportional with pressure: Thus, its value will be different at different positions along the column. In eqs.11.18 and 11.20 we are using the average linear carrier gas velocity; however, in general, we do not calculate D_M for an average pressure. Therefore, one may want to express the Van Deemter-Golay equation to include the value of D_{Mo} and the diffusion coefficient at outlet pressure, and carry out the modifications necessary to consider the pressure changes along the column. The resulting general equation — based on the work of Giddings et al. [12,13] and written according to Cramers et al. [14] — is:

$$H = A \cdot j" + \frac{B \cdot j"}{u_o} + C_M \cdot j" \cdot u_o + C_S \cdot j \cdot u_o \quad \text{eq.11.41}$$

In eq.11.41, u_o is the mobile phase velocity at column outlet, j is the gas compression correction factor (Section 5.3) and $j"$ is the pressure correction factor of Giddings (Section 5.4.2); the four basic terms (A, B, C_M and C_S) are equal to those described earlier:

packed columns:	A	term:	eq.11.22
	B	term:	eq.11.23
	C_M	term:	eq.11.24
	C_S	term:	eq.11.25
open-tubular columns:	A	term:	it is equal to zero
	B	term:	eq.11.28
	C_M	term:	eq.11.29
	C_S	term:	eq.11.30 or 11.31

As mentioned, D_{Mo}, the diffusion coefficient in the gaseous mobile phase calculated at outlet pressure, must be used in the calculation of the B and C_M terms for both types of columns.

11.4.7 Reduced Plate Height Equation (Knox Equation)

This relationship was first described by Knox [15, 16] utilizing the dimensionless terms of the reduced plate height *(h)* and reduced mobile phase velocity *(v)*:

$$h = H / d_p \qquad\qquad \text{eq.7.15a}$$

$$v = \overline{u} \cdot d_p / D_M \qquad\qquad \text{eq.5.11}$$

where d_p is the particle diameter, D_M is the diffusion coefficient of the analyte in the mobile phase, H is the HETP and $\overline{u}$ is the average linear mobile phase velocity. Eqs.7.15a and 5.11 are written for packed columns; in the case of open-tubular columns the inner tube diameter *(d_c)* is to be used instead of the particle diameter. The *reduced plate height equation* is an empirical relationship written in the following form:

$$h = A \cdot v^{1/3} + \frac{B}{v} + C \cdot v \qquad \text{eq.11.42}$$

where the meaning of the A, B and C terms is the same as discussed earlier in connection with eq.11.18; however, their numerical values are different. Eq.11.42 is commonly called the *Knox equation*. Figure 22 shows the logarithmic plot corresponding to eq.11.42 (the usual presentation), together with the contributions of the individual terms.

The advantage of the reduced plate height equation is that the values of the individual terms are independent of the type of column, analyte or chromatographic technique. According to Knox, for a fairly good column the values of the three constants are: $A = 1$, $B = 2$ and $C = 0.1$.

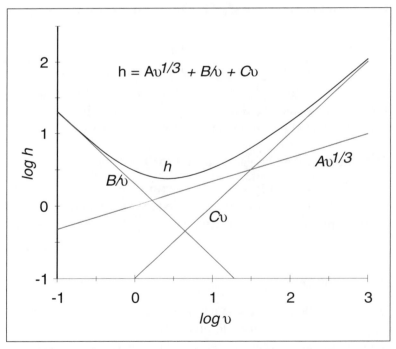

Figure 22. Logarithmic plot of h *(reduced plate height) vs.* v *(reduced mobile phase velocity) according to the Knox equation, indicating also the contributions of the three individual terms of the equation.*

11.5 Retention Time vs. Other Parameters

The total retention time of a peak (t_R) was expressed earlier (see Section 3.2) in the following way:

$$t_R = t_M (1+k) \qquad\qquad \text{eq.3.5}$$

We have also seen that (Section 5.2.2):

$$t_R = \frac{L}{\overline{u}}(1+k) \qquad\qquad \text{eq.5.10}$$

where t_M is the mobile phase holdup time, L is the column length, k is the retention factor (capacity ratio) of the analyte and $\overline{u}$ is the average linear velocity of the mobile phase.

Since the column length that is necessary to achieve a given resolution depends on the number of theoretical plates required to separate a certain peak pair, we are particularly interested in how the fundamental chromatographic parameters influence the time of analysis. In this respect "time of analysis" refers to the retention time of the second peak in a peak pair characterized by a relative retention (α) and retention factor (k_2), which is separated by resolution R_s. If we express L from eq.11.8:

$$L = 16 R_s^2 \left(\frac{\alpha}{\alpha - 1} \right)^2 \left(\frac{k_2 + 1}{k_2} \right)^2 H_2 \qquad\qquad \text{eq.11.43}$$

and substitute it into eq.5.10, the following fundamental relationship is obtained [17]:

$$t_R = 16 R_s^2 \left(\frac{\alpha}{\alpha - 1} \right)^2 \left[\frac{(1 + k_2)^3}{k_2^2} \cdot \frac{H_2}{\overline{u}} \right] \qquad\qquad \text{eq.11.44}$$

In this equation the influence of the two terms in the brackets is particularly important.

If we investigate the variation of the $(1+k)^3/k^2$ term with k (see Figure 23, page 132), we can observe that it has a minimum at $k = 2$. This means that one should try — whenever possible — to bring the capacity ratio of the peaks representing the most critical separation (by the proper selection of the stationary phase, column temperature and column type) close to this value.

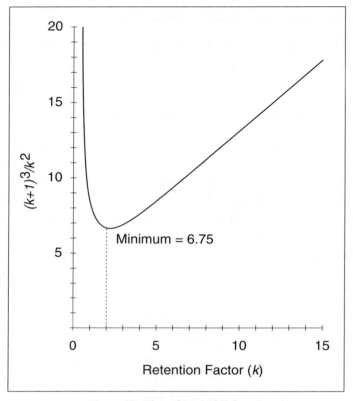

Figure 23. Plot of $(1+k)^3/k^2$ *vs.* k

If we compare the second term in the brackets with the Van Deemter-Golay plots in Figure 20 on page 116, it is clear that it is equal to the C term of the Van Deemter-Golay equation, i.e., it gives the slope of the linear plot approaching the ascending part of the HETP vs. $\bar{u}$ curve. We have already indicated that generally the C term is smaller for open-tubular columns than for packed columns. This means that, in general, the same resolution can be obtained in a shorter time on an open-tubular column than on a packed column.

11.6 Comparison of Chromatographic Techniques Through the Reduced Parameters

In this book we have discussed the basic relationships used

in gas chromatography. However, as pointed out in the fore-word, most of these can be adapted well to other column chro-matographic techniques.

At the end of our book, we would like to briefly compare a few aspects of the three — gas, supercritical-fluid and liquid — column chromatographic techniques. This can be done through the reduced terms. Their advantage is exactly this: They permit the comparison of different types of columns and chromato-graphic techniques. The following three reduced terms are of interest in this respect:

Reduced mobile phase velocity:

$$v = \frac{\overline{u} \cdot d_c}{D_M}$$

eq.5.12

Reduced plate height:

$$h = H / d_c$$

eq.7.15b

Reduced film thickness:

$$\delta_f = \frac{d_f}{d_c}\sqrt{\frac{D_M}{D_S}}$$

eq.4.16

The fourth reduced term used here is the retention factor (capacity ratio: k) and its variations (see Part III):

$$k = K / \beta$$

eq.11.45

In these equations K is the distribution constant (partition coefficient), β is the phase ratio of the column, d_c is the inner column tube diameter, d_f is the liquid phase film thickness, H is the height equivalent to one theoretical plate (HETP), $\overline{u}$ is the average linear mobile phase velocity, and D_M and D_S are the diffusion coefficients of the analyte in the mobile and station-ary phases, respectively.

Above, the reduced terms v and h were written for open-tubular columns. The same equations are also valid for packed columns, only in such a case d_p, the particle diameter, should replace d_c, the inner tube diameter.

In Section 11.1 we have already dealt with the influence of the retention factor (k) and the interpretation of eq.11.45. With respect to the present discussion it is important that when comparing different techniques, the comparison should be carried out on a constant k basis.

Our comparison of the different column chromatographic techniques will be based on the speed of analysis and the inner column tube diameter vs. the coated stationary phase film thickness. Finally, some questions related to efficiency and velocity will also be investigated. Here we are considering open-tubular columns; however, with the proper adjustments, the same treatments are also applicable to packed columns.

11.6.1 Speed of Analysis

We have seen (cf. eqs.3.5 and 5.10) that the total retention time can be described by the following relationship:

$$t_R = t_M (1+k) = \frac{L}{\overline{u}}(1+k)$$

Substituting $L = N \cdot H$ (cf.eq.7.3) and then expressing $\overline{u}$ from eq.5.12 and H from eq.7.15b:

$$t_R = \frac{N \cdot H}{\overline{u}}(1+k) \qquad \text{eq.11.46}$$

$$\overline{u} = \frac{v \cdot D_M}{d_c} \qquad \text{eq.11.47}$$

$$H = h \cdot d_c \qquad \text{eq.11.48}$$

Substituting these equations for $\overline{u}$ and H into eq.11.46:

$$t_R = N \frac{h}{v} \cdot \frac{d_c^2}{D_M}(1+k) \qquad \text{eq.11.49}$$

This means that under identical conditions $(N, h, v$ and $k)$, the analysis time is proportional to $(d_c)^2/D_M$ [or $(d_p)^2/D_M$ for packed columns]. The order of magnitude of D_M in the various techniques is [18]:

gas chromatography (GC): 10^{-1} cm²/s

supercritical-fluid chromatography (SFC): 10^{-4} cm²/s

liquid chromatography (LC): 10^{-5} cm²/s

It can be established from the above given relationships that, in order to get equal analysis time, the diameter of open-tubular columns used in the three chromatographic techniques should be in the ratio of 1.0 : 0.3 : 0.01 (ref. [18]). In other words, if the inner diameter of the open-tubular column used in GC is 250 μm, then it should be 75 μm in SFC and 2.5 μm in LC. If packed columns are used in LC (because an open-tubular column with an internal diameter of 2.5 μm is impractical), then the particle diameter of the column packing should be around 2.5 μm.

11.6.2 Film Thickness

In the previous section we have indicated the diameters of open-tubular columns needed in the various chromatographic techniques to achieve equal analysis time. An additional question is the corresponding liquid phase film thickness, and this can be evaluated with the help of the reduced film thickness (δ_f) which gives the d_f/d_c ratio:

$$\frac{d_f}{d_c} = \delta_f \sqrt{\frac{D_S}{D_M}}$$

eq.11.50

where d_c is the inner tube diameter, d_f is the stationary phase film thickness and D_S and D_M are the diffusion coefficients of the analyte in the stationary and mobile phases, respectively. As pointed out by Schoenmakers, who introduced this term [19], the value of δ_f should remain unchanged, independent of the chromatographic technique. If the same stationary phase (e.g., a siloxane polymer) is used in the various chromatographic techniques, then D_S remains the same, and thus the d_c/d_f ratio can be adjusted according to the change in the $\sqrt{D_M}$ value.

Eq.11.50 means that if diffusion in the mobile phase is slowed down (as in SFC and LC as compared to GC), then the stationary phase film can be relatively thicker (i.e., the phase ratio of the column smaller). Assuming the same value for δ_f and D_S, we can write eq.11.50 for two different chromatographic techniques (different D_M values):

$$\frac{d_{f1}}{d_{f2}} \cdot \frac{d_{c2}}{d_{c1}} = \sqrt{\frac{D_{M2}}{D_{M1}}}$$

and

$$d_{f2} = d_{f1} \cdot \frac{d_{c2}}{d_{c1}} \sqrt{\frac{D_{M1}}{D_{M2}}} \qquad \text{eq.11.51}$$

Naturally, the tube diameter has to be reduced as discussed above, in order to maintain a comparable speed of analysis. If we consider the tube diameters and the orders of magnitude of D_M as given in Section 11.6.1, the following approximate film thickness values should be used; as interesting information, the corresponding phase ratio (β) values are also given:

GC:	d_c =	250	μm	d_f = 0.25 μm	β =	250
SFC:	d_c =	75	μm	d_f = 2.0 μm	β =	9.4
LC:	d_c =	2.5	μm	d_f = 0.25 μm	β =	2.5

11.6.3 Efficiency

Literature data [20] based on liquid chromatographic measurements with packed columns indicate a reduced plate height value at optimum of about $h = 2$. On the other hand, Schoenmakers [19] gives values of about 0.8–1.3 for an analyte having a retention factor of $k = 3$. According to eq.11.48 this would mean that the measured $HETP_{min}$ values should be about 0.8–1.3 times the inner diameter of an open-tubular column, regardless of the particular chromatographic technique. This range corresponds fairly well to our earlier considerations of gas chromatography. The theoretical value of $HETP_{min}$ was described in eq.11.36b as:

$$HETP_{min(theor)} = d_c \sqrt{\frac{1 + 6k + 11k^2}{12(1+k)^2}} \qquad \text{eq.11.36b}$$

The value of the square root function varies between 0.29 for $k = 0$ and about unity for $k = \infty$ (see Table XXI, page 121); for $k = 3$ it is equal to 0.78. In other words, $HETP_{min(theor)}$ at $k = 3$ should correspond to $0.78 \cdot d_c$. If we assume a 60% utilization of the theoretical efficiency (see Section 11.4.4) — which is about the worst case — then the actual HETP would correspond to $1.3 \cdot d_c$. This corresponds well to the range given by Schoenmakers.

11.6.4 Velocity

Schoenmakers [19] gave reduced velocity values of about 3–5 at optimum for an analyte with $k = 3$, and literature data[21] based on LC measurements are also in this range. This would indicate (cf. eq.11.47) that the average linear gas velocity at optimum ($\overline{u}_{opt}$) is about 3–5 times the D_M / d_c ratio. Turning to gas chromatography, if we consider the D_M values listed in Table XXII on page 124 for n-heptadecane and a 0.25 mm I.D. open-tubular column, a factor of three would give the respective $\overline{u}_{opt}$ values of about 11, 37 and 46 cm/s for nitrogen, helium and hydrogen as the carrier gas, while the corresponding values with a factor of five would be 19, 61 and 77 cm/s. These latter values are close to the calculated $\overline{u}_{opt(theor)}$ values (see Table XXII); however, these cases significantly differ from the actually measured $\overline{u}_{opt}$ values.

This brief discussion shows that while in the case of $HETP_{min}$, both the theoretical considerations and conclusions drawn from the reduced values are not far from actual values, one must be careful in drawing conclusions concerning $\overline{u}_{opt}$ from calculated values, either from the simplified theoretical relationship or from reduced values.

11.7 References

[1] J.H. Purnell, *J. Chem. Soc.* **1960**, 1268–1274.

[2] J.H. Knox, *J. Chem. Soc.* **1961**, 433–441.

[3] H.A.C. Thijssen, *J. Chromatogr.* **11**, 141–150 (1963).

[4] A.S. Said, *J. Gas Chromatogr.* **2**, 60–71 (1964).

[5] J.J. Van Deemter, F.J. Zuiderweg and A. Klinkenberg, *Chem. Eng. Sci.* 5, 271–289 (1956).

[6] M.J. Golay, in *Gas Chromatography 1958 (Amsterdam Symposium)* (D.H. Desty, ed.), Butterworths, London, 1958; pp. 36–55.

[7] K.J. Hyver ed., *High-Resolution Gas Chromatography*. 3rd ed., Hewlett-Packard Co., Avondale, PA 1989; p.1–16.

[8] D.H. Desty, A. Goldup and D.H.F. Whyman, *J. Inst. Petrol.* 45, 287–298 (1959).

[9] L.S. Ettre in *Gas Chromatography 1966 (Rome Symposium)* (A.B. Littlewood, ed.), Inst. of Petroleum, London, 1967; pp.115–118.

[10] M.J.E. Golay, *Anal. Chem.* 40, 382–384 (1968).

[11] L.S. Ettre and J.E. Purcell, in *Advances in Chromatography, Vol.10* (J.C. Giddings and R.A. Keller, eds.), M. Dekker, Inc., New York, 1974; pp. 1–97.

[12] G.H. Stewart, S.L. Seager and J.C. Giddings, *Anal. Chem.* 31, 1738 only (1959).

[13] J.C. Giddings, S.L. Seager, L.R. Stucki and G.H. Stewart, *Anal. Chem.* 32, 867–870 (1960).

[14] C.A. Cramers, C.E. Van Tilburg, C.P.M. Schutjes, J.A. Rijks, G.A. Rutten and R. De Nijs, in *Capillary Chromatography (5th Symposium, Riva del Garda, April 26–28, 1983)* (J. Rijks, ed.), Elsevier, Amsterdam, 1983; pp. 76–84.

[15] P.A. Bristow and J.H. Knox, *Chromatographia* 10, 279–289 (1977).

[16] J.H. Knox, *J. Chromatogr. Sci.* 15, 353–364 (1977).

[17] L.S. Ettre and E.W. March, *J. Chromatogr.* 91, 5–24 (1974).

[18] H.G. Janssen and C.A. Cramers, *J. Chromatogr.* 505, 19–35 (1990).

[19] P.J. Schoenmakers, *JHRC/CC* 11, 278–282 (1988).

[20] P.A. Bristow, *LC in Practice*. HETP Inc., Wilmslow, Cheshire, 1976; p.19.

[21] V.R. Meyer, *Praxis der Hochleistungsflüssigchromatographie*. 6th ed. Salle/Sauerländer Verlag, Frankfurt/Main, 1990; p.101.

Supplements

I

Derivation of the Relationship Used to Calculate the (Gas) Holdup Time from the Retention Times of Three Members of a Homologous Series

This derivation is based on the linear relationship between the logarithm of the adjusted retention time (t_R') and the carbon number (c_n) of members of a homologous series:

$$\log t_R' = a \cdot c_n + b \qquad\qquad \text{eq.12.1}$$

where a and b are constants. If we write this equation for two consecutive homologs having the carbon numbers of z and ($z+1$):

$$\log t_{R(z+1)}' = a\,(z+1) + b$$

$$\log t_{Rz}' = a \cdot z + b$$

and then subtract the second equation from the first, we obtain that:

$$\log t_{R(z+1)}' - \log t_{Rz}' = a(z+1) - a \cdot z = a \qquad\qquad \text{eq.12.2}$$

or

$$\log\!\left[t_{R(z+1)}' \,/\, t_{Rz}' \right] = a \qquad\qquad \text{eq.12.3}$$

But the bracketed term is the relative retention (r) of the two consecutive homologs. If the logarithm of a value is constant, then the value is also a constant; therefore, we can write that:

$$r_{(z+1)/z} = const. \qquad \text{eq.12.4}$$

Naturally, this relationship is valid not only for two consecutive homologs, but also for any homologous pairs that are equidistant in their carbon number.

"Equidistant" means that if subindices *1, 2* and *3* refer to three homologs with increasing retention time ($t_{R1} < t_{R2} < t_{R3}$), then:

$$c_{n2} = c_{n1} = c_{n3} - c_{n2} \qquad \text{eq.12.5}$$

In such a case:

$$r_{2/1} = r_{3/2} \qquad \text{eq.12.6a}$$

or

$$t'_{R2}/t'_{R1} = t'_{R3}/t'_{R2} \qquad \text{eq.12.6b}$$

Let us now substitute ($t_R - t_M$) for t_R' and then carry out the corresponding modifications of the equation:

$$\frac{t_{R2} - t_M}{t_{R1} - t_M} = \frac{t_{R3} - t_M}{t_{R2} - t_M}$$

$$(t_{R2} - t_M)(t_{R2} - t_M) = (t_{R1} - t_M)(t_{R3} - t_M)$$

$$t_{R2}^2 - 2t_M \cdot t_{R2} + t_M^2 = t_{R1} \cdot t_{R3} - t_M \cdot t_{R3} - t_M \cdot t_{R1} + t_M^2$$

$$t_{R2}^2 - 2t_M \cdot t_{R2} = t_{R1} \cdot t_{R3} - t_M \cdot t_{R3} - t_M \cdot t_{R1}$$

$$t_M \cdot t_{R3} - t_M \cdot t_{R2} + t_M \cdot t_{R1} - t_M \cdot t_{R2} = t_{R1} \cdot t_{R3} - t_{R2}^2$$

Reorganizing the last equation we get:

$$t_M(t_{R3} - t_{R2}) - t_M(t_{R2} - t_{R1}) = t_{R1} \cdot t_{R3} - t_{R2}^2 \qquad \text{eq.12.7}$$

Let us use the following symbols:

$$x_1 = t_{R2} - t_{R1} \qquad \text{eq.12.8a}$$

$$x_3 = t_{R3} - t_{R2} \qquad \text{eq.12.8b}$$

Substituting these symbols into eq.12.7 and expressing t_M, we obtain:

$$t_M = \frac{t_{R1} \cdot t_{R3} - t_{R2}^2}{x_3 - x_1}$$ eq.12.9

which is one version of the equation used to calculate the (gas) holdup time from the retention times of three homologs with equidistant carbon numbers.

Equation 12.9 can be simplified. First we express t_{R1} from eq.12.8a and t_{R3} from eq.12.8b:

$$t_{R1} = t_{R2} - x_1$$
$$t_{R3} = t_{R2} + x_3$$

We substitute these expressions for t_{R1} and t_{R3} in the numerator of eq.12.9:

$$t_{R1} \cdot t_{R3} - t_{R2}^2 = (t_{R2} - x_1)(t_{R2} + x^3) =$$
$$t_{R2}^2 + t_{R2} \cdot x_3 - t_{R2} \cdot x_1 - x_1 \cdot x_3 - t_{R2}^2 =$$
$$t_{R2}(x_3 - x_1) - x_1 \cdot x_3$$

Resubstituting this for the numerator of eq.12.9 we obtain:

$$t_M = \frac{t_{R2}(x_3 - x_1) - x_1 \cdot x_3}{x_3 - x_1} = t_{R2} - \frac{x_1 \cdot x_3}{x_3 - x_1}$$ eq.12.10

This is the modified equation to calculate the (gas) holdup time.

II

Calculation of the Film Thickness from Static Coating Data

The coating solution completely fills the column tube, having an inside radius of r_c and a length of L; thus, the volume of the coating solution in the tube is:

$$V_{tube} = r_c^2 \pi \cdot L$$

Within this volume, the volume of the liquid phase is V_L, and the volume of the solvent is V_{solv}. The concentration of the liquid phase in the coating solution (in wt/vol %) is c_L and the density of the liquid phase is ρ (g/mL). We can express the volumes of the liquid phase and of the solvent in the following way:

$$V_L = \frac{\left(r_c^2 \pi \cdot L\right)c_L}{100\rho} = \frac{\left(r_c^2 \pi \cdot L\right)(c_L / \rho)}{100} \qquad \text{eq.12.11}$$

$$V_{solv} = r_c^2 \pi \cdot L - \frac{(r_c^2 \pi \cdot L)(c_L / \rho)}{100} = (r_c^2 \pi \cdot L)\frac{100 - (c_L / \rho)}{100} \qquad \text{eq.12.12}$$

The ratio of the two volumes is:

$$\frac{V_L}{V_{solv}} = \frac{c_L / \rho}{100 - \left(c_L / \rho\right)} \qquad \text{eq.12.13}$$

After the evaporation of the solvent from the column, V_L will correspond to the volume of the liquid phase in the final coated column, and V_{solv} will be equal to V_G, the volume of the gas phase in the column. Hence:

$$\frac{V_L}{V_G} = \frac{V_L}{V_{solv}} = \frac{c_L/\rho}{100-(c_L/\rho)} \qquad \text{eq.12.14}$$

But $V_L/V_G = 1/\beta$, the inverse of the phase ratio of the column. If $r_c \gg d_f$, then $\beta = r_c/2d_f$ (see Section 4.8). Hence:

$$\frac{2d_f}{r_c} = \frac{c_L/\rho}{100-(c_L/\rho)} \qquad \text{eq.12.15}$$

From eq.12.15 we can express either d_f or c_L:

$$d_f = \frac{r_c}{2}\left[\frac{(c_L/\rho)}{100-(c_L/\rho)}\right] \qquad \text{eq.12.16}$$

$$c_L = \frac{(200d_f)\cdot\rho}{r_c+2d_f} \qquad \text{eq.12.17}$$

Table XXIII on page 146 lists density values for the most common silicone phases.

Code **	Chemical characterization	ρ 25°/25°	wt.av. mol.wt.
OV-1	Dimethyl silicone gum	0.980	>10^6
OV-101	Dimethyl silicone fluid	0.975	30,000
OV-225	Methyl (50%) cyanopropyl (25%) phenyl (25%) silicone	1.096	8,000
Methyl phenyl silicones			
OV-73	5.5% Phenyl	0.991	800,000
OV-3	10 % Phenyl	0.997	20,000
OV-7	20 % Phenyl	1.021	10,000
OV-61	33 % Phenyl	1.090	40,000
OV-11	35 % Phenyl	1.057	7,000
OV-17	50 % Phenyl	1.092	4,000
OV-22	65 % Phenyl	1.127	8,000
OV-25	75 % Phenyl	1.150	10,000

*Table XXIII. Data of selected silicone phases **

---

* From L.S. Ettre, *Chromatographia* **34**, 513–528 (1992). The data are from Catalogue No.39 (1990) of Ohio Valley Specialty Chemical Co., Marietta, OH. For an explanation of the meaning of the chemical composition, see the footnote to Table XIX on page 97.

** Code numbers of Ohio Valley Specialty Chemical Co.

III

Derivation of the Three Versions of the Fundamental Relationship Between Efficiency, Resolution, Selectivity and Retention

First we modify the equation describing the number of theoretical plates (N): t_R and t_R' are the total and adjusted retention times, respectively; w_b is the peak width at base; and k is the retention factor (capacity ratio):

$$N = 16\left(\frac{t_R}{w_b}\right)^2 = 16\left(\frac{t_R'+t_M}{w_b}\right)^2 = 16\left(\frac{t_R'}{w_b}\right)^2\left(\frac{t_R'+t_M}{t_R'}\right)^2$$

$$\frac{t_R'+t_M}{t_R'} = \frac{t_R'/t_M+t_M/t_M}{t_R'/t_M} = \frac{k+1}{k}$$

Thus:

$$\frac{N}{16} = \left(\frac{t_R'}{w_b}\right)^2\left(\frac{k+1}{k}\right)^2 \qquad\qquad \text{eq.12.18}$$

$$\frac{t_R'}{w_b} = \frac{\sqrt{N}}{4}\cdot\frac{k}{k+1} \qquad\qquad \text{eq.12.19}$$

First Version

We use the simplified equation for resolution (R_s) (cf. eq.8.11):

$$R_s = (t'_{R2} - t'_{R1}) / w_{b2} \qquad \text{eq.12.20}$$

Modifying eq. 12.20:

$$R_s = \frac{t'_{R2} - t'_{R1}}{w_{b2}} = \frac{t'_{R2} - t'_{R1}}{t'_{R2}} \cdot \frac{t'_{R2}}{w_{b2}} \qquad \text{eq.12.21}$$

We now modify the first fraction on the right-hand side of eq.12.21 by dividing both the numerator and denominator by t'_{R1}:

$$\frac{t'_{R2} - t'_{R1}}{t'_{R2}} = \frac{t'_{R2} / t'_{R1} - t'_{R1} / t'_{R1}}{t'_{R2} / t'_{R1}} = \frac{\alpha - 1}{\alpha} \qquad \text{eq.12.22a}$$

Resubstituting this into eq.12.21 and substituting the right-hand side of eq.12.19 for t'_{R2} / w_{b2}:

$$R_s = \frac{\sqrt{N_2}}{4} \cdot \frac{\alpha - 1}{\alpha} \cdot \frac{k_2}{k_2 + 1} \qquad \text{eq.12.22b}$$

Expressing N_2 (now called N_{req}):

$$N_{req} = 16 R_s^2 \left(\frac{\alpha - 1}{\alpha} \right)^2 \left(\frac{k_2 + 1}{k_2} \right)^2 \qquad \text{eq.12.23}$$

Second Version

Now we write the simplified resolution equation for the first peak:

$$R_s = \left(t'_{R2} - t'_{R1} \right) / w_{b1} \qquad \text{eq.12.24}$$

We modify this equation in the same way as in the First Version, but now using t'_{R1}. The resulting relationships are:

$$R_s = \frac{t'_{R2} - t'_{R1}}{t'_{R1}} \cdot \frac{t'_{R1}}{w_{b1}}$$ eq.12.25

$$\frac{t'_{R2} - t'_{R1}}{t'_{R1}} = \frac{t'_{R2}/t'_{R1} - t'_{R1}/t'_{R1}}{t'_{R1}/t'_{R1}} = \alpha - 1$$ eq.12.26

$$\frac{t'_{R1}}{w_{b1}} = \frac{\sqrt{N}}{4} \cdot \frac{k_1}{k_1 + 1}$$ eq.12.27

Resubstituting these expressions into eq.12.25:

$$R_s = \frac{\sqrt{N_1}}{4} (\alpha - 1) \frac{k_1}{k_1 + 1}$$ eq.12.28

Expressing N_1 (now called N_{req}) from eq.12.28:

$$N_{req} = 16 R_s^2 \left(\frac{1}{\alpha - 1} \right)^2 \left(\frac{k_1 + 1}{k_1} \right)^2$$ eq.12.29

Third Version

Here the average values already defined in eqs.11.12–11.14 are used; plate number N^* is expressed as given in eq.11.15. We start the derivation by expressing resolution according to eq.8.10:

$$R_s = \frac{t'_{R2} - t'_{R1}}{\dfrac{w_{b1} + w_{b2}}{2}} = \frac{t'_{R2} - t'_{R1}}{\overline{w_b}}$$

This equation is further modified by multiplying the right-hand side by $\overline{t'_R}/\overline{t'_R}$:

$$R_s = \left(\frac{t'_{R2} - t'_{R1}}{\overline{t'_R}} \right) \left(\frac{\overline{t'_R}}{\overline{w_b}} \right)$$ eq.12.30

$$(a) \qquad (b)$$

For the further modification of (a) we express $\overline{t'_R}$ from eq.11.12 and then divide both the numerator and the denominator by t'_{R1} :

$$\frac{t'_{R2}-t'_{R1}}{\overline{t'_R}} = \frac{2(t'_{R2}-t'_{R1})}{t'_{R1}+t'_{R2}} =$$

$$\frac{2[t'_{R2}/t'_{R1}-t'_{R1}/t'_{R1}]}{t'_{R1}/t'_{R1}+t'_{R2}/t'_{R1}} = \frac{2(\alpha-1)}{\alpha+1}$$

eq.12.31

For (b) we substitute the proper term derived earlier in eq.12.19, but now written for N^* :

$$\frac{\overline{t'_R}}{\overline{w_b}} = \frac{\sqrt{N^*}}{4} \cdot \frac{\overline{k}}{\overline{k}+1}$$

eq.12.32

Reconstructing eq.12.30 we obtain:

$$R_s = \frac{\sqrt{N^*}}{2} \cdot \frac{\alpha-1}{\alpha+1} \cdot \frac{\overline{k}}{\overline{k}+1}$$

eq.12.33

Expressing N^* from eq.12.33:

$$N^* = 4R_s^2 \left(\frac{\alpha+1}{\alpha-1}\right)^2 \left(\frac{\overline{k}+1}{\overline{k}}\right)^2$$

eq.12.34

IV

Books on Gas Chromatography

As already emphasized in the Foreword, our book is not a general textbook on gas chromatography. Our aim is to summarize and explain the most important terms, definitions and relationships in gas chromatography; to provide an aid in the interpretation of chromatographic results; and to help in the understanding of more complicated texts. For a detailed study of the underlying principles, theory and practice of (gas) chromatography, our readers are advised to consult the existing general and specialized textbooks. There are many, and in this final part of our book we try to present a selective listing.

Chromatography has a unified theory, common with the general theory of most separation techniques. Therefore, our listing starts by giving a few books dealing with separation techniques and chromatography in general. These are then followed by general textbooks on gas chromatography and by books dealing with specialized topics. Books discussing the theory and practice of open-tubular columns are combined in a separate subchapter. This is followed by books dealing with the application of gas chromatography in various fields and by books discussing the combination of gas chromatography with other major instrumental techniques. Finally, we list the existing chromatography handbooks and mention the book series of the individual publishing houses specializing in chromatography.

Our listing includes only books published in English; those that were originally published in another language are marked with (T). Symposium proceedings are not included in the listing. The listing is chronological; where we know of more than one edition, this is indicated. The size of the books (number of pages) is always indicated, as well as the authors' or editors' names, the publisher and the year of publication. Some books have been published by two publishers, one in Europe and another in the U.S.A. A full list of the publishers' addresses is given at the end of this part. It should be noted that in the more than 35 years spanned by our listing, a number of changes have occurred in the field of book publication: some publishers ceased to exist or merged with others, and the name of a publisher may be listed in different ways at various times. We tried to list them as uniformly as possible.

Most of the books included in this listing are still available: to use the term of the publishing field, they are *in print*. We also included some basic books which are now out of print and cannot be purchased anymore; however, most major libraries and the libraries of industrial companies using chromatography in their laboratories have at least some of them on their shelves. These books are marked with an asterisk (*). These books are not "obsolete"; in fact, often they present the fundamentals of (gas) chromatography in even more detail than the more recent books, which probably concentrate more on the technique. Of course, there are a number of other books that have been published in the past but are not included in our listing: In those cases we felt that more recent publications superseded the earlier books.

This listing is based on the publishers' catalogues and our own book collection. A few books might have been excluded by oversight; still, we believe that this exhaustive listing of 108 titles is representative of the books dealing with various aspects of gas chromatography.

I. Books on Separation Techniques in General

The bulk of these books deal with the theory of chromatography as one of the most important separation techniques.

B.L. KARGER, L.R. SNYDER and C. HORVÁTH: *Introduction to Separation Science.* (*) Wiley, 1973; 586 pp.

J.M. MILLER: *Separation Methods in Chemical Analysis.* (*) Wiley, 1975; 309 pp.

J.C. GIDDINGS: *Unified Separation Science.* Wiley, 1991; 320 pp.

2. General Textbooks on Chromatography

These books summarize the common theory of chromatography and then deal separately with the individual chromatographic techniques, discussing their most important aspects.

J.C. GIDDINGS: *Dynamics of Chromatography. Principles and Theory.* Dekker, 1965: 2nd reprint; 323 pp.

A.S. SAID: *Theory and Mathematics of Chromatography.* Huethig, 1981; 210 pp.

C.F. POOLE and S.A. SCHUETTE: *Contemporary Practice of Chromatography.* Elsevier, 1984: 2nd reprint, 1986; 708 pp.

P.A. SEWELL and B. CLARKE: *Chromatographic Separation.* Wiley, 1987; 335 pp.

R. KALISZAN: *Quantitative Structure — Chromatographic Retention Relationships.* Wiley, 1987; 303 pp.

J.M. MILLER: *Chromatography — Concepts and Contrasts.* Wiley, 1988; 297 pp.

R.M. SMITH: *Gas and Liquid Chromatography in Analytical Chemistry.* Wiley, 1988; 402 pp.

C.F. POOLE and S.K. POOLE: *Chromatography Today.* Elsevier, 1991; 1026 pp.

E. HEFTMANN (ed.): *Chromatography:* Vol.I: *Fundamentals and Techniques;* Vol.II: *Applications.* 5th ed. Elsevier, 1992; Vol.I, 552 pp., Vol.II, 630 pp.

3. General Textbooks on Gas Chromatography

3a Fundamental Books Published Before 1970

These books are now out of print (indicated by the asterisk); however, they are still available in larger libraries. These are the fundamental books from which tens of thousands have learned the principles and practice of gas chromatography. Books marked with (T) represent the English translations of books originally published in another language.

C.G.S. PHILLIPS: *Gas Chromatography.* (*) Butterworths-Academic Press, 1956; 105 pp.

A.I.M. KEULEMANS: *Gas Chromatography.* (*) Chapman & Hall — Reinhold; 1st ed. 1957, 217 pp.; 2nd ed. 1959, 234 pp.

E. BAYER: *Gas Chromatography.* (*)(T) Elsevier, 1961; 240 pp.

D. AMBROSE and B. AMBROSE: *Gas Chromatography.* (*) George Newnes, London, 1961 — Van Nostrand, 1962; 220 pp.

J.H. KNOX: *Gas Chromatography.* (*) Methuen & Co. — Wiley, 1962; 126 pp.

A.B. LITTLEWOOD: *Gas Chromatography: Principles, Techniques and Applications.* (*) Academic Press; 1st ed. 1962, 507 pp.; 2nd ed. 1970, 546 pp.

S. DAL NOGARE and R.S. JUVET Jr.: *Gas Liquid Chromatography: Theory and Practice.* (*) Wiley, 1962; 450 pp.

J.H. PURNELL: *Gas Chromatography.* (*) Wiley, 1962; 441 pp.

L.S. ETTRE and A. ZLATKIS (eds.): *The Practice of Gas Chromatography.* (*) Wiley, 1967; 591 pp.

O.E. SCHUPP III: *Gas Chromatography.* (*) Wiley, 1968; 437 pp.

J. TRANCHANT (ed.): *Practical Manual of Gas Chromatography.* (*)(T) Elsevier, 1969; 387 pp.

3b Textbooks Published Since 1980

These books are still in print and are available from the publishers.

R.L. GROB (ed.): *Modern Practice of Gas Chromatography*. Wiley, 1st ed. 1977, 654 pp.; 2nd ed. 1985, 897 pp.

J.A. PERRY: *Introduction to Analytical Gas Chromatography: History, Principles and Practice*. Dekker, 1981; 448 pp.

J. WILLETT: *Gas Chromatography*. Wiley, 1987; 253 pp.

W. JENNINGS: *Analytical Gas Chromatography*. Academic Press, 1987; 259 pp.

G. GUIOCHON and C.L. GUILLEMIN: *Quantitative Gas Chromatography for Laboratory and On-Line Process Control*. Elsevier, 1988; 798 pp.

G. SCHOMBURG: *Gas Chromatography: A Practical Course*. VCI Publishers, 1990; 320 pp.

4. Specialized Techniques of Gas Chromatography

In this group we present books on specialized techniques. The first four books are out of print; these are, however, included because they represent the most thorough discussion of their subject.

W.E. HARRIS and H.W. HABGOOD: *Programmed Temperature Gas Chromatography*. (*) Wiley, 1966; 305 pp.

V.G. BEREZKIN: *Analytical Reaction Gas Chromatography*. (*)(T) Plenum, 1968; 193 pp.

D.A. LEATHARD and B.C. SHURLOCK: *Identification Techniques in Gas Chromatography*. (*) Wiley, 1970; 282 pp.

A. ZLATKIS and V. PRETORIUS (ed.): *Preparative Gas Chromatography*. (*) Wiley, 1971; 402 pp.

V.G. BEREZKIN (ed.): *Chemical Methods in Gas Chromatography*. (T) Elsevier, 1983; 314 pp.

A. KATSANOS: *Flow Perturbation Gas Chromatography*. Dekker, 1988; 320 pp.

D.R. LLOYD, T.C. WARD, H.P. SCHREIBER and C.C. PIZANA (eds.): *Inverse Gas Chromatography*. A.C.S., 1989; 331 pp.

5. Books on Specialized Topics in Gas Chromatography

The books listed here deal with specialized topics in gas chromatography and with separate parts of the GC system.

5a Theory of Gas Chromatography

J. NOVÁK: *Quantitative Analysis by Gas Chromatography.* Dekker; 1st ed. 1975, 218 pp.; 2nd ed. 1988, 360 pp.

R. VILCU and M. LECA: *Polymer Thermodynamics by Gas Chromatography.* Elsevier, 1990; 204 pp.

5b Gas Adsorption Chromatography

The book by Kiselev and Yashin (now out of print) is the most thorough discussion of the theory of gas adsorption chromatography and the adsorbents used as stationary phases.

A.V. KISELEV and Ya.I. YASHIN: *Gas Adsorption Chromatography.* (*)(T) Plenum, 1969; 254 pp.

T. PARYJCZAK: *Gas Chromatography in Adsorption and Catalysis.* Ellis Horwood, 1986; 346 pp.

V.G. BEREZKIN: *Gas-Liquid-Solid Chromatography.* (T) Dekker, 1991; 312 pp.

5c Sample Handling and Sampling

The book by Blau and King represents the most detailed discussion of all the possible derivatives used in all types of chromatographic techniques. The book is now out of print.

K. BLAU and G.S. KING: *Handbook of Derivatives for Chromatography.* (*) Heiden, 1977; 576 pp.

J. DROZD: *Chemical Derivatization in Gas Chromatography.* Elsevier, 1981; 232 pp.

W.G. JENNINGS and A. RAPP: *Sample Preparation for Gas Chromatographic Analysis.* Huethig, 1983; 104 pp.

5d Headspace Sampling

Headspace sampling for gas chromatography is an important technique for the analysis of volatile compounds in a non-volatile matrix. The instrumentation part of the Hachenberg-Schmidt book is now fairly obsolete; however, the special part dealing with the application of HS-GC for physico-chemical measurements is still important. The Kolb book is a collection of special articles. Both books, originally published by Heyden & Sons, continue to be marketed by Wiley.

H. HACHENBERG and A.P. SCHMIDT: *Gas Chromatographic Headspace Analysis.* (T) Heyden, 1977; 1st reprint 1979; 2nd reprint by Wiley, 1983; 125 pp.

B. KOLB (ed.): *Applied Headspace Gas Chromatography.* Heyden, 1980; 185 pp.

B.V. IOFFE and A.G. VITENBERG: *Head-Space Analysis and Related Methods in Gas Chromatography.* (T) Wiley, 1984; 276 pp.

5e Pyrolysis-Gas Chromatography

Pyrolysis-gas chromatography had been an important technique for the investigation of non-volatile compounds, essentially polymers.

R.W. MAY, E.F. PEARSON and D. SCOTHERN: *Pyrolysis Gas Chromatography.* Chemical Society, 1977; 109 pp.

S.A. LIEBMAN and E.J. LEVY: *Pyrolysis and Gas Chromatography in Polymer Analysis.* Dekker, 1985; 576 pp.

5f Stationary Phases

Rotzsche's book is a very detailed discussion of practically every stationary phase used since the inception of gas chromatography (also including adsorbents). The first 42 pages of the book present a concise summary of the principles and theory of GC. The book by König deals with a very important new group of compounds used as stationary phases and with their applications for enantiomer separation.

H. ROTZSCHE: *Stationary Phases in Gas Chromatography*. Elsevier, 1991; 410 pp.

W.A. KÖNIG: *Gas Chromatographic Enantiomer Separation with Modified Cyclodextrins*. Huethig, 1992; 168 pp.

5g Detectors

The development in the field of GC detectors in the past 20 years has been considerable; however, although new selective detectors have been added to the classical detectors, the principles remained the same. In order to present a complete listing, two out-of-print books are also listed.

D.J. DAVID: *Gas Chromatographic Detectors*. (*) Wiley, 1974; 295 pp.

J. SEVCIK: *Detectors in Gas Chromatography*. (*) Elsevier, 1976; 192 pp.

M. DRESSLER: *Selective Gas Chromatographic Detectors*. Elsevier, 1986; 192 pp.

H.H. HILL and D.G. McMINN (eds.): *Detectors for Capillary Chromatography*. Wiley, 1992; 444 pp.

5h Troubleshooting of GC Systems

These two books specialize in the maintenance and troubleshooting of gas chromatographic systems.

J.Q. WALKER, M.T. JACKSON, Jr. and J.B. MAYNARD: *Chromatographic Systems: Maintenance and Troubleshooting*. Academic Press, 1st ed. 1972; 2nd ed. 1977, 359 pp.

D. ROOD: *A Practical Guide to the Care, Maintenance and Troubleshooting of Capillary Gas Chromatographic Systems*. (**T**) Huethig, 1991; 191 pp.

6. Open-Tubular Column Gas Chromatography

The first book in the listing by Ettre — which is now out of print — presents the first concise summary of the theory and practice of open-tubular column gas chromatography but, obviously, it did not deal with glass or fused-silica columns and with the more recent stationary phases.

A book on troubleshooting of open-tubular column systems was listed earlier, while other books dealing with the applications of open-tubular columns and with the combination of open-tubular column gas chromatography with liquid chromatography are included in the next sections.

L.S. ETTRE: *Open-Tubular Columns in Gas Chromatography.* (*) Plenum, 1965; 164 pp.

W. JENNINGS: *Gas Chromatography with Glass Capillary Columns.* Academic Press, 1st ed. 1978, 184 pp.; 2nd ed. 1980, 320 pp.

W.G. JENNINGS: *Comparison of Fused-Silica and Other Glass Columns in Gas Chromatography.* Huethig, 1981; 81 pp.

M.L. LEE, F.J. YANG and K.D. BARTLE: *Open-Tubular Column Gas Chromatography: Theory and Practice.* Wiley, 1984; 445 pp.

P. SANDRA (ed.): *Sample Introduction in Capillary Gas Chromatography.* Huethig, 1985; 265 pp.

K. GROB: *Classical Split and Splitless Injection in Capillary Gas Chromatography.* Huethig, 1986; 324 pp.

K. GROB: *Making and Manipulating Capillary Columns for Gas Chromatography.* Huethig, 1986; 250 pp.

W.A. KÖNIG: *The Practice of Enantiomer Separation by Capillary Gas Chromatography.* Huethig, 1987; 120 pp.

K. GROB: *On-Column Injection in Capillary Gas Chromatography.* Huethig, 1987, 2nd printing 1991; 591 pp.

A. VAN ES: *High Speed, Narrow-Bore Capillary Gas Chromatography.* Huethig, 1992; 143 pp.

7. Books on the Application of Gas Chromatography

7a Environmental Analysis

R.L. GROB and M.A. KAISER: *Environmental Problem Solving Using Gas and Liquid Chromatography.* Elsevier, 1982, 1st reprint 1985; 240 pp.

R.L. GROB: *Chromatographic Analysis of the Environment.* Dekker, 1983 (2nd ed.); 736 pp.

F.L. ONUSKA and F.W. KARASEK: *Open-Tubular Gas Chromatography in Environmental Sciences.* Plenum, 1984; 296 pp.

V.G. BEREZKIN and Yu.S. DRUGOV: *Gas Chromatography in Air Pollution Analysis.* (T) Elsevier, 1991; 212 pp.

7b Trace Analysis

These two books are now out of print. We still list them here because they represent the first thorough discussion of this special field.

H. HACHENBERG: *Industrial Gas Chromatographic Trace Analysis.* (*) Heyden, 1973; 217 pp.

V.G. BEREZKIN and V.S. TATARINSKII: *Gas Chromatographic Analysis of Trace Impurities.* (*)(T) Plenum - Consultant Bureau, 1973; 177 pp.

7c Gas Analysis

The first book was published in the first decade of modern gas chromatography; still, it is one of the best discussions of this important and complex application.

P.G. JEFFERY and P.J. KIPPING: *Gas Analysis by Gas Chromatography.* (*) Pergamon, 1964; 213 pp.

C.J. COWPER and A.J. DEROSE: *The Analysis of Gases by Gas Chromatography.* Pergamon, 1983; 159 pp.

7d Analysis of Essential Oils

W. JENNINGS and T. SHIBAMOTO: *Quantitative Analysis of Flavor and Fragrance Volatiles by Glass Capillary Gas Chromatography.* Academic Press, 1980; 472 pp.

P. SANDRA and C. BICCHI (eds.): *Capillary Gas Chromatography in Essential Oil Analysis.* Huethig, 1987; 435 pp.

7e Biochemical, Clinical and Medical Analysis

B.M. MITRUKA (ed.): *Gas Chromatographic Applications in Microbiology and Medicine.* Wiley, 1975; 472 pp.

H. JAEGER: *Glass Capillary Chromatography in Clinical Medicine and Pharmacology.* Dekker, 1985; 656 pp.

R.W. ZUMWALT, K.C. KUO and C.W. GEHRKE: *Amino Acid Analysis by Gas Chromatography.* CRC Press, 1987; 3 volumes, 630 pp.

A.R. McCAFFERTY and D. WILSON (eds.): *Chromatography and Isolation of Insect Hormones and Pheromones.* Plenum, 1990; 390 pp.

R.E. CLEMENT (ed.): *Gas Chromatography: Biochemical, Biomedical and Clinical Applications.* Wiley, 1990; 393 pp.

T.A. GOUCH: *The Analysis of Drugs of Abuse.* Wiley, 1991; 628 pp.

7f Various Applications

Five books that could not be included in the previous listings are given here.

J.R. CONDER and C.L. YOUNG: *Physico-Chemical Measurements by Gas Chromatography.* Wiley, 1979; 632 pp.

K.H. ALTGELT and T.H. GOUW: *Chromatography in Petroleum Analysis.* Dekker, 1979; 512 pp.

W. JENNINGS (ed.): *Applications of Glass Capillary Gas Chromatography.* Dekker, 1981; 629 pp.

R.W. SOUTER: *Chromatographic Separations of Stereoisomers.* CRC Press, 1985; 256 pp.

W.G. JENNINGS and J.G. NIKELLY (eds.): *Capillary Chromatography — The Applications.* Huethig, 1991; 153 pp.

8. Books on Hyphenated Techniques

This term has been used recently to describe the direct (online) combination of two or more major instrumental techniques. Books listed below deal with the coupling of gas chromatography with mass spectrometry, Fourier-transform infrared spectroscopy, atomic emission spectroscopy and liquid chromatography.

G.M. MESSAGE: *Practical Aspects of Gas Chromatography — Mass Spectrometry*. Wiley, 1984; 351 pp.

G. ODHAM, L. LARSSON and P.A. MÅRDH: *Gas Chromatography — Mass Spectroscopy: Applications in Microbiology*. Plenum, 1984; 460 pp.

W. HERRES: *Capillary Gas Chromatography — Fourier Transform Infrared Spectroscopy*. Huethig, 1987; 210 pp.

F.W. KARASEK and R.E. CLEMENT: *Basic Gas Chromatography — Mass Spectrometry*. Elsevier, 1988 (1st reprint 1991); 202 pp.

K. GROB: *On-Line Coupled LC-GC*. Huethig, 1991; 462 pp.

P.C. UDEN (ed.): *Element Specific Chromatographic Detection by Atomic Emission Spectroscopy*. ACS, 1992; 345 pp.

9. Handbooks of Chromatography

9a CRC Handbook Series

This book series is published by CRC Press. Besides general summaries, these books mainly consist of tabulated data on columns, retention times, etc. The books listed here contain data obtained by GC, LC and TLC. The series also includes other books; however, those do not contain GC data.

G. ZWEIG and J. SHERMA (eds.): *General Data and Principles*. 1973; Vol.I, 784 pp.; Vol.II 343 pp.

S.C. CHURMS (ed.): *Carbohydrates*. 1981; 288 pp.

R.N. GUPTA and I. SHUNSHINE (eds.): *Drugs*. 1981; Vol.I, 352 pp.; Vol.II, 416 pp.

R.N. GUPTA (ed.): *Drugs*. 1988; Vol.III, 272 pp.; Vol.IV, 448 pp.; Vol.V, 384 pp., Vol.VI, 448 pp.

T. HANAI (ed.): *Phenols and Organic Acids*. 1982; 304 pp.

G. ZWEIG, J. SHERMA, C.G. SMITH, N.E. SKELLY, C.D. CHOW and R. SOLOMON (eds.): *Polymers*. 1982; 200 pp.

S. BLACKBURN (ed.): *Amino Acids and Amines*. Vol.I, 1983: 312 pp.; Vol.II, 1989: 432 pp.

J.M. FELLWEILER and J. SHERMA: *Pesticides and Related Compounds*. 1984; 376 pp.

H.K. MANGOLD (ed.): *Lipids*. 1984; Vol.I, 624 pp.; Vol.II, 368 pp.

M. QURESHI (ed.): *Inorganics*. 1986; 384 pp.

S. BLACKBURN (ed.): *Peptides*. 1986; 400 pp.

J.C. TOUCHSTONE (ed.): *Steroids*. 1986; 272 pp.

W.L. ZIELINSKI (ed.): *Hydrocarbons by Gas Chromatography*. 1987; 248 pp.

9b Dictionary

There exists a very useful, four-language (English-German-French-Russian) dictionary of the chromatographic terms. The first edition was published in Eastern Germany and distributed in the western world by Pergamon Press, while its second edition was published by Huethig:

H-P. ANGELÉ: *Technical Dictionary of Chromatography*. Verlag Technik, Berlin — Pergamon, 1970; 119 pp.

H-P ANGELÉ: *Dictionary of Chromatography*. Huethig, 1984; 141 pp.

10. Book Series

Four publishers have book series specially devoted to chromatography:
- Elsevier has the *Journal of Chromatography Library* series. As of this date 53 volumes have been published in this series.

- Dekker has the *Chromatographic Science* series. As of this date, 56 volumes have been published in this series.

- Huethig has the *Chromatographic Method* series. As of this date close to 40 volumes have been published in this series.

- CRC Press is publishing the *Handbooks of Chromatography* from which the pertinent volumes were listed above, in Section 9a.

Many of the books listed by us here have been published as part of these series. In addition, most publishers also have special series devoted to analytical chemistry, and these also include books on chromatography and its applications.

In our listing, we disregarded the special volumes representing *symposium proceedings*. Particularly in the first two decades of the development of modern chromatography, these proceedings included very important original reports on the newest advances of the techniques. While today most of the proceedings are published as regular volumes of certain journals, up to about the mid-1970s, they were published as regular books, which can be found in major libraries.

11. List of Publishers

Academic Press: Academic Press
1250 Sixth Avenue
San Diego, CA 92101-4311

A.C.S.: American Chemical Society
Books Department
1155 Sixteenth Street NW
Washington, DC 20036

Chapman & Hall: Chapman & Hall Ltd.
11 New Fetter Lane
London, EC4P 4EE, United Kingdom

Chemical Society: The Royal Society of Chemistry
Burlington House
Picadilly
London, W1V 0BN, United Kingdom

CRC Press: CRC Press Inc.
2000 Corporate Boulevard NW
Boca Raton, FL 33431

Dekker:	Marcel Dekker, Inc. 270 Madison Avenue New York, NY 10016
Ellis Horwood:	Ellis Horwood Ltd. Market Cross House, Cooper Street Chichester, West Sussex PO19 1EB United Kingdom
Elsevier:	Elsevier Science Publishers P.O. Box 330 1000 AH Amsterdam The Netherlands
	Elsevier Science Publishing Co. Inc. P.O. Box 882 Madison Square Station New York, NY 10159
Heyden:	The former Heyden & Son publishers (London, U.K.) merged with John Wiley & Sons.
Huethig:	Dr. Alfred Huethig Verlag GmbH P.O. Box 102869 D-6900 Heidelberg 1, Germany
Pergamon:	Pergamon Press Inc. Hiddington Hill Hall Oxford OX3 OBW United Kingdom
	Pergamon Press, Inc. 660 White Plains Road Tarrytown, NY 10591-5153
Plenum:	Plenum Publishing Corporation 233 Spring Street New York, NY 10013-1578
Reinhold:	Van Nostrand Reinhold Co. 450 West 33rd Street New York, NY 10001

Van Nostrand:	Van Nostrand Reinhold Co.
	450 West 33rd Street
	New York, NY 10001
VCI Publishers:	VCI Publishers
	220 East 23rd Street, Suite 909
	New York, NY 10010-4606
Wiley:	John Wiley & Sons
	605 Third Avenue
	New York, NY 10158-0012

List of Symbols

General Information

The rules and recommendations of the Division of Physical Chemistry of the International Union of Pure and Applied Chemistry (I.U.P.A.C.) [1] prescribe the following general symbols for major physical and physico-chemical quantities and units:

A	area
d	diameter
D	diffusion coefficient
k	rate constant
K	equilibrium constant
p, P	pressure
r	radius
t	time
T	temperature (degrees Kelvin)
u	velocity
V	volume
W	mass (weight)
ρ	density
η	viscosity

The Commission on Chromatography and Other Analytical Separations of the Analytical Chemistry Division of I.U.P.A.C. recently accepted a new and comprehensive Nomenclature for Chromatography [2]. This nomenclature is utilizing these basic symbols, with three additions:

F volumetric flow rate

L length

w peak width

and a differentiation between p (for pressure) and P (for relative pressure). Further differentiation is made by the utilization of superscripts and subscripts, based on the following general rules:

- In general, composite symbols are to be avoided.

- **Superscripts** are used for adjusted ($'$) and corrected (^{o}) retention times and volumes, and to specifically indicate data obtained under programmed-temperature conditions (^{T}).

- Physical conditions or the phase are indicated by **capitalized subscripts** such as, e.g., M and S for the mobile and stationary phase, respectively, or in gas chromatography, G for gas and L for the liquid phase. Thus, e.g., the diffusion coefficient in the mobile phase is D_M and not D_m.

  In addition, **capitalized subscripts** are used for "retention" (R: thus, the retention time is t_R and not t_r) and for "net" (N as in t_N and V_N, the net retention time and volume, respectively).

- Physical parts of the system are generally characterized by **lower-case subscripts**, such as, e.g., c for "column," p for "particles" and f for "film."

- To indicate a given compound if there is already a subscript, one should never use double subscripts; rather, the additional subscript should be added to the existing subscript. If the existing or additional subscripts consist of more than one character or number, then the additional information should be given in parentheses.

Thus, while it is t_{Ri} or t_{R1}, it should be $t_{R(st)}$ or $t_{R(z+1)}$; however, it should never be $t_{R_{st}}$ or $t_{R_{z+1}}$.

- Subscript o generally refers to outlet of a column, but is also used in a number of terms to describe some base values.

In this book, we have utilized the symbols included in the new I.U.P.A.C. chromatography nomenclature and also followed these rules in further differentiations.

List of Symbols Used

The symbols and acronyms used in this book are listed below. The section where the symbol is defined or first used is always indicated.

Symbols

a	constant (in various equations)
a	front part of peak width (2.5)
a	factor expressing the thickness of the porous layer in a PLOT (SCOT) column, as a fraction of the tube radius (11.4.5)
a_1	factor expressing the tortuous path in the porous layer of a PLOT (SCOT) column, as a fraction of the tube radius (11.4.5)
a_2	factor expressing the thickness of the average free gas volume in the porous layer of a PLOT (SCOT) column, as a fraction of the tube radius (11.4.5)
a_K	factor in sample capacity expression (10.1)
a'	constant (in various equations)
A	constant (in various equations)
A	"eddy diffusion" term in the Van Deemter equation (11.4)
A	integrated peak area (2.4)
A_{coated}	actually coated surface in a column (4.5)
A_{geom}	geometric area of column's inner surface (4.1)
A'	constant (in various equations)
A'	peak area calculated as $m_{max} \cdot w_h$ (2.4)
A_c	cross-sectional area of a column available to the carrier gas flow (4.2)

As	asymmetry factor of a peak (2.5)
b	constant (in various equations)
b	back part of peak width (2.5)
b'	constant (in various equations)
B	constant (in various equations)
B	longitudinal diffusion term in the Van Deemter-Golay equations (11.4)
B'	constant (in various equations)
B_o	specific permeability (4.10)
c	concentration of an analyte in the detector (10.3.1)
c_L	concentration of the liquid phase in a coating solution (wt/vol %) (4.5)
c_n	number of carbon atoms in a molecule (of a member of a homologous series) (2.2)
C	resistance to mass transfer term in the Van Deemter-Golay equations (11.4)
C^*	value proportional to sample capacity of a column (10.1)
C_M	term describing the resistance to mass transfer in the mobile phase, used in the Van Deemter-Golay equations (11.4)
C_S	term describing the resistance to mass transfer in the stationary phase, used in the Van Deemter-Golay equations (11.4)
d_c	column tube inside diameter (Part IV)
d_f	liquid phase film thickness (4.5)
d_p	particle diameter (Part IV)
d_{pl}	thickness of the porous layer in a PLOT (SCOT) column (11.4.5)
D_{AB}	diffusion coefficient of analyte A in mobile phase B (5.6)
D_M	diffusion coefficient in the mobile phase (5.6)
D_{Mo}	diffusion coefficient in the mobile phase, at column outlet pressure (5.6, 11.4.6)
D_S	diffusion coefficient in the stationary phase (4.12)
f	front part of peak width (2.5)
f	relative detector response factor (10.6)

f	general symbol for various functions of k (Table XXI)
F	factor in the original Golay equation, expressing the increase in the coated area relative to the geometric inside area of an open-tubular column (11.4.2)
F	mobile phase flow rate measured at column outlet and ambient temperature, using a wet flow meter (5.1)
F	value used in the calculation of separation power (8.5)
F_a	mobile phase flow rate measured at column outlet and ambient temperature, corrected to dry gas conditions (5.1)
F_c	mobile phase flow rate measured at column outlet and corrected to column temperature (5.1)
G	value used in the calculation of separation power (8.5)
h	reduced plate height (7.4)
H	plate height (height equivalent to one theoretical plate; *HETP*) (7.2)
H^*	plate height corresponding to N^* (11.3.3)
H_{min}	minimum plate height of the Van Deemter-Golay plots ($HETP_{min}$) (11.4)
$H_{min(theor)}$	theoretically calculated best value of HETP (the theoretical minimum of the Van Deemter-Golay plots [$HETP_{min(theor)}$] (11.4.3)
H_{eff}	effective plate height (height equivalent to one effective plate) (7.3)
ΔH_s	heat of solution (6.7)
i	difference in the retention index values of two analytes on the same column (9.3)
I	retention index; Kováts retention index (9.1)
ΔI	difference in the retention index values of the same analyte, measured on a polar vs. a non-polar phase (9.6)
I^T	linear retention index; retention index obtained in programmed-temperature analysis (9.4)

j	mobile phase compression (compressibility) correction factor (5.3)
j'	pressure correction factor of Halász (5.4.1)
j''	pressure correction factor of Giddings (5.4.2)
$k\ (k')$	retention factor (capacity ratio) (3.1)
$\bar{k}$	average retention factor (11.3.3)
K	distribution constant (partition coefficient) (1.3)
L	column length (Part IV)
m	peak height in general, at any position on the peak (2.3)
m_{max}	peak height at maximum (2.3)
M	molecular mass (weight) (5.6)
n	number of peaks to be placed between two major peaks (8.4)
N	plate number (number of theoretical plates) (7.1)
$\bar{N}$	average plate number (the mean of the plate numbers calculated for two peaks) (11.3.3)
N^*	plate number calculated from average retention time and peak width values (11.3.3)
N_{eff}	effective plate number (number of effective plates) (7.3)
N_{req}	number of theoretical plates required to achieve a given resolution (11.3)
$\bar{p}$	average pressure (5.8.1)
p^o	saturated vapor pressure of the analyte (1.3)
Δp	pressure drop (5.5)
p_a	ambient pressure (5.1)
p_i	inlet pressure (5.5)
p_o	outlet pressure (5.5)
p_z	pressure at point z in the column (5.8.1)
p_w	partial pressure of water at ambient temperature (5.1)
P	relative pressure for a column (ratio of inlet and outlet pressures) (5.2.1)
P_z	relative pressure at point z in a column (ratio of pressure p_z and the outlet pressure) (5.8.1)
P	general polarity of a stationary phase (9.6.3)
$\bar{P}$	mean polarity of a stationary phase (9.6.3)

q	configuration factor in the C_S term of the Van Deemter equation for packed columns (11.4.1)
r	relative retention (8.1.2)
r_c	column tube inner radius (Part IV)
r_G	unadjusted relative retention (8.2)
R	retardation factor (3.5)
R	gas constant (1.3; 4.12; 6.7)
R_s	peak resolution (8.3)
s	one of the Rohrschneider constants (9.6.1)
s'	one of the McReynolds constants (9.6.2)
S	detector sensitivity (10.4)
S_{sup}	specific surface area of the support (4.5)
t_M	mobile phase hold-up time; retention time of an unretained compound (2.1)
t_N	net retention time (2.1; 6.6)
t_R	total retention time (2.1)
$\bar{t}_R$	average total retention time (11.3.3)
$t_R{}^T$	total retention time in programmed-temperature analysis (9.4)
t_R'	adjusted retention time (2.1)
$t_R{}^\circ$	corrected retention time (2.1; 6.4)
Δt	expression for $(t_{R2} - t_{R1})$ or $(t_{R2}' - t_{R1}')$ (8.3)
T_a	ambient temperature (5.1)
T_c	column temperature (1.3)
T_R	retention temperature (9.4)
u	mobile phase velocity (5.2)
u	one of the Rohrschneider constants (9.6.1)
$\bar{u}$	average linear carrier gas velocity (5.2.2)
u'	one of the McReynolds constants (9.6.2)
u_i	carrier gas velocity at column inlet (5.8)
u_o	carrier gas velocity at column outlet (5.2.1)
$\bar{u}_{opt}$	average linear carrier gas velocity corresponding to the minimum of the Van Deemter-Golay plots (11.4)
$\bar{u}_{opt(theor)}$	theoretically calculated optimum average carrier gas velocity (the value of the average velocity corresponding to the theoretical minimum of the Van Deemter-Golay plots) (11.4)

u_z	mobile phase velocity at point z in the column (5.8.2)
v	volume of analyte's vapor at column temperature (10.3.1)
Σv	sum of atomic volume increments (Fuller-Schettler-Giddings equation) (5.6.1)
v_{eff}	effective volume of one plate (10.1)
V_A	molar volume of the analyte (Wilke-Chang equation) (5.6.2)
V_c	column tube's inner volume (4.1)
V_e	extra-column ("dead") volumes (6.2)
V_g	specific retention volume at 0°C (6.7)
V_g^{θ}	specific retention volume at column temperature (6.7)
V_G	volume of gas phase in the column (4.3)
V_L	volume of liquid phase in the column (4.5)
V_M	volume of mobile phase in the column (1.3; 4.3)
V_M	mobile phase hold-up volume; retention volume of an unretained compound (6.1)
V_M°	corrected gas hold-up volume (6.2)
V_{max}	limiting value of analyte vapor to be introduced in a column (10.1)
V_N	net retention volume (6.6)
V_o	interparticle volume of column (4.2)
V_p	peak volume (mobile phase volume corresponding to a peak) (2.6)
V_R	total retention volume (6.3)
V_R'	adjusted retention volume (6.5)
V_R°	corrected retention volume (6.4)
V_S	volume of stationary phase in the column (1.3)
w	peak width at any point (2.3)
w_b	peak width at base (2.3)
$\overline{w}_b$	average peak width at base (11.3.3)
w_h	peak width at half height (2.3)
w_i	peak width at inflection points (2.3)
Δw	absolute difference $\mid b - f \mid$ of peak width (2.5)
$W(W_i)$	amount (weight) of compound i present (1.3)

$W_{i(M)}$	amount (weight) of compound i present in the mobile phase (1.3)
$W_{i(S)}$	amount (weight) of compound i present in the stationary phase (1.3)
W_L	amount (weight) of liquid phase in the column (4.7)
$W_L\%$	liquid phase loading of a packed column (4.7)
W_{max}	upper limit of linearity of a detector (amount, concentration or mass flow of the analyte) (10.2)
W_{min}	minimum detectability of a detector (amount, concentration or mass flow of the analyte) (10.2)
W_t	mass (amount) of analyte entering a detector in unit time (10.3.2)
W_S	amount (weight) of stationary phase in the column (6.7)
W_{sup}	amount (weight) of support in the column (4.5)
x	difference in retention times (Suppl. No. I)
x	one of the Rohrschneider constants (9.6.1)
x'	one of the McReynolds constants (9.6.2)
y	one of the Rohrschneider constants (9.6.1)
y'	one of the McReynolds constants (9.6.2)
z	number of carbon atoms of an n-alkane eluting before the peak of interest (9.1)
z	one of the Rohrschneider constants (9.6.1)
z'	one of the McReynolds constants (9.6.2)
$z+1$	number of carbon atoms of an n-alkane eluting after the peak of interest (9.1)

Greek letters

α	separation factor (relative retention) (8.1)
α_G	unadjusted separation factor (unadjusted relative retention) (8.2)
β	phase ratio of a column (4.8)
γ	tortuosity factor in the B term of the Van Deemter equation for packed columns (11.4.1)
γ°	activity coefficient of the analyte at infinite dilution in the stationary phase (1.3)
δ	separation power (8.5)
δ_f	reduced liquid phase film thickness (4.6)

ε	interparticle porosity (4.2)
η	mobile phase viscosity (5.7)
ϕ	a function representing the ratio of peak height at a given point to the peak height at maximum (2.3)
λ	packing factor in the A term of the Van Deemter equation for packed columns (11.4.1)
υ	reduced mobile phase viscosity (5.2.3)
ρ	density of stationary (liquid) phase at column temperature (1.3:4.5)
σ	standard deviation of a Gaussian peak (2.3)
σ^2	variance of a Gaussian peak (2.3)
Φ	flow resistance parameter of a column (4.11)
Ψ_B	association factor (solvent constant) in the Wilke-Chang equation (5.6.2)
ω	packing factor in the C_M term of the Van Deemter equation for packed columns (11.4.1)

Acronyms

FID	flame-ionization detector (1.2)
EPN	effective peak number (8.4.1)
GC	gas chromatography (1.1)
HETP	height equivalent to one theoretical plate (7.2)
HETP$_{min}$	HETP value corresponding to the minimum of the Van Deemter-Golay plots (11.4)
HETP$_{min(theor)}$	theoretically calculated best value of HETP (the theoretical minimum of the Van Deemter-Golay plots) (11.4.3)
HWD	hot-wire detector (1.2)
IR	infrared (spectrophotometer) (1.1)
LC	liquid chromatography (1.1)
LR	linear range (of a detector) (10.2)
PLOT	porous-layer open-tubular (column) (1.2)
PN	peak number (8.4)
RRT	relative retention time (unadjusted relative retention) (8.2)
SCOT	support-coated open-tubular (column) (1.2)
SFC	supercritical-fluid chromatography (1.1)
SN	separation number (Trennzahl) (8.4.2)

TCD	thermal-conductivity detector (1.2)
TZ	Trennzahl (separation number) (8.4.2)
UTE%	utilization of theoretical efficiency (11.4.4)
WCOT	wall-coated open-tubular (column) (1.2)

References

[1] Manual of Symbols and Terminology for Physicochemical Quantities and Units. *Pure Appl. Chem.* **21**, 3–44 (1977).

Nomenclature for Chromatography. *Pure Appl. Chem.* **65**, 819–872 (1993).